INTEGRATED POLLUTION PREVENTION AND CONTROL FOR THE MUNICIPAL WATER CYCLE IN A RIVER BASIN CONTEXT

VALIDATION OF THE THREE-STEP STRATEGIC APPROACH

T0176440

ALBERTO GALVIS CASTAÑO

Thesis committee

Promotor
Prof. Dr H.J. Gijzen
Professor of Environmental Biotechnology
IHE Delft Institute for Water Education
Regional Director and Representative
UNESCO Regional Office for Southern Africa
Harare, Zimbabwe

Co-promotor
Dr N. P. van der Steen
Associate Professor of Environmental Technology
IHE Delft Institute for Water Education

Other members
Prof. Dr C. Kroeze, Wageningen University & Research
Prof. Dr C.J. van Leeuwen, Utrecht University
Prof. Dr M. von Sperling, Federal University of Mina Gerais, Brazil
Prof. Dr P. van der Zaag, IHE Delft, TU Delft

Integrated Pollution Prevention and Control for the Municipal Water Cycle in a River Basin Context

Validation of the Three-Step Strategic Approach

Thesis
submitted in fulfilment of the requirements of
the Academic Board of Wageningen University and
the Academic Board of the IHE Delft Institute for Water Education
for the degree of doctor
to be defended in public
on Tuesday, 3 September 3, 2019 at 3:00 p.m.
in Delft, the Netherlands

by

Alberto Galvis Castaño
Born in Sevilla-Valle del Cauca, Colombia

CRC Press/Balkema is an imprint of the Taylor & Francis Group, an informa business

© 2019, Alberto Galvis Castaño

Although all care is taken to ensure integrity and the quality of this publication and the information herein, no responsibility is assumed by the publishers, the author nor IHE Delft for any damage to the property or persons as a result of operation or use of this publication and/or the information contained herein.

A pdf version of this work will be made available as Open Access via http://repository.tudelft.nl/ihe. This version is licensed under the Creative Commons Attribution-Non Commercial 4.0 International License, http://creativecommons.org/licenses/by-nc/4.0/

Published by:
CRC Press/Balkema
Schipholweg 107C, 2316 XC, Leiden, the Netherlands
Pub.NL@taylorandfrancis.com
www.crcpress.com – www.taylorandfrancis.com

ISBN: 978-0-367-37527-0 (Taylor & Francis Group)
ISBN: 978-94-6343-951-0 (Wageningen University & Research)
DOI: https://doi.org/10.18174/475836

*To my wife Nohelia and
to my sons Juan Manuel and Juan José*

Acknowledgements

The idea of developing this research started with the beginning of the new millennium, when I was fortunate to work together with my promotor, Professor Huub Gijzen. This occurred within the framework of the cooperation between IHE and Universidad del Valle, with support of the Netherlands Government. During this cooperation, we had the opportunity to learn about some experiences in Colombia and Latin America in urban water management related to water quality recovery of rivers for their different uses. Many of these experiences were associated with significant investments and, paradoxically, with low impact, as the quality of water resources kept deteriorating. In this context, Professor Gijzen encouraged me to get involved in a doctoral program that would allow me to explore the problem in more depth and to try to contribute to finding solutions.

I wish to thank my promotor Professor Huub Gijzen from UNESCO and my co-promotor Dr. Peter van der Steen, from IHE-Delft, for their guidance, coaching, support and patience through the development of doctoral research. I am grateful to them for their rigor in reviewing all the chapters in this document. This process has been very important in my academic training, my role as a lecturer at the Universidad del Valle and as Director of the Research Group on Integrated Water Resource Management.

I am also grateful for the valuable support of colleagues at the Cinara Institute of the Faculty of Engineering of the Universidad del Valle, especially its Director Luis Darío Sánchez. I also thank the support of my five students from Universidad del Valle: Diana Alexandra Zambrano, María Fernanda Jaramillo, Isabel Cristina Hurtado, Faber Montaña and Juan Gabriel Urrego.

This research would not have been possible without the support of different institutions among others in providing access to information. These include: Corporación Autónoma Regional del Valle del Cauca (CVC); Empresas Municipales de Cali (Emcali) and Regional Cleaner Production Center (CRPML). I also want acknowledge my colleagues and friends: Manfred Schütze, Amparo Duque, Gloria Almario, Ana Dorly Jaramillo, Luis German Delgado, Javier Ernesto Holguín and Carlos Arturo Martínez. With these colleagues I had the opportunity to analyse and discuss several of the key topics of this research, both personally and through projects in which we had the opportunity to work together. I would like to extend my gratitude to my friend and colleague Rubén Dario Pinzon, expert designer, builder and operator of wastewater treatment plants both in Colombia and in other Latin American countries who helped me in the process of analysing the aspects of my PhD research related to wastewater treatment plant.

I would also like to thank Jan Teun Visccher, who has had a close relationship with the Cinara Institute since its inception. With Jan Teun, I had the opportunity to work on the SWITCH project, on the Learning Alliances topic, having Cali as a demonstration city. I am very grateful to Jan Teun for his support and hospitality during my visits to Delft, in the development of my PhD.

A special thanks to my brother and colleague Gerardo Galvis, founder of Cinara Institute. During the 1980s and 1990s he led a process of change in the Universidad del Valle, related to postgraduate academic programs and the research and development projects in the Sanitary and Environmental Engineering field. This change contributed to the consolidation of research groups and the training at the doctoral level of its members.

Last but not least a very special thanks to my lovely family: my wife Nohelia, my sons Juan Manuel and Juan Jose, for their love, motivation and unconditional support during my research.

Summary

The trend towards more urbanized societies and the growing number of people has significant implications for freshwater use and wastewater management. Factors such as climate change are making the problems related to water even more critical. At the same time, water resources are also being substantially affected by human activities such as dam building, deforestation, erosion, mining activities, land use changes and pollutant load discharges. In many cases, especially in developing countries, the protection of water resources from quality deterioration by point and non-point source pollution has been based on 'end-of-pipe' solutions. That strategy only considers, in terms of infrastructure, wastewater treatment plants (WWTPs). Often this is accompanied by adjustments to the regulations, including the application of economic instruments, such as taxes for wastewater discharge. However, this strategy has not completely fulfilled the objective of recovering the quality of the water resource for its different uses. Continuing the urban water practice in a 'business-as-usual' manner is unsustainable, considering its implications for public health, environment and, thus, the economy. Water should be administrated as a limited resource with multiple uses, and any solutions should be formulated with appropriate distribution and protection criteria, considering the basin as a planning unit. To face this situation and the challenges of the Sustainable Development Goals (SDGs), a systematic vision is necessary to guarantee the effectiveness of investments in water and sanitation. The 2030 Agenda recognizes the centrality of water resources for sustainable development and the vital role that improved drinking water, sanitation and hygiene play in progress in other areas, including health, education and poverty reduction.

The research described in this thesis intends to contribute to the solution of the previously outlined problem. In this research a technology-selection approach to control pollution by domestic wastewater was investigated. The technology selection involved multi-criteria analysis (MCA), the application of mathematical modelling of water quality in rivers, and cost-benefit analysis (CBA). The basin was used as a unit of analysis and the technical, environmental, social, cultural, economic, policy and regulatory aspects were considered integrally. This research was oriented towards the validation of a strategy of technology selection based on the Three-Step Strategic Approach concept (3-SSA). In this context, 'technology selection' will not be understood as merely the treatment technology, but it includes such aspects as minimisation and prevention, both in the urban water cycle (housing and urban drainage system) and interventions at the basin level, WWTPs, reuse of effluents, and the natural and/or stimulated self-purification capacity of the water bodies.

In this doctoral research each step of the 3-SSA was studied independently. The results and conclusions of the study of each step were an input to perform the comprehensive analysis of the sequential implementation (chronological order) of the three steps. The 3-SSA (Unconventional strategy) was validated by applying it to the Upper Cauca river basin (La Balsa-Anacaro Stretch: 389 km) in Colombia. This study included the comparison with a Conventional Strategy, which considered a 'business-as-usual scenario' of high water use, 'end-of-pipe' wastewater treatment and conventional water supply providing drinking water quality for all uses. The implementation of the 3-SSA included a reduction in water consumption (Step1) and reuse of treated wastewater in households and for sugarcane crop irrigation (Step 2). It also considered the prioritization of investments to maximize the impact in improving the water quality of the Cauca River, targeting interventions in watersheds and municipalities with the highest pollutant load and located upstream of the river segments with the lowest DO (Step 3). The CBA clearly favoured the 3-SSA (Unconventional Strategy). The 3-SSA led to a major impact on the water quality of the Cauca River. This result was mainly due to the large differences in initial investment and O&M costs of WWTP in municipalities for the two strategies. For the Unconventional Strategy the WWTPs were smaller due to the application of the prevention and minimization approaches and treatment for reuse. The application of the 3-SSA resulted in avoided costs for initial investments and O&M, especially for groundwater wells and associated pumps for sugarcane crop irrigation. Furthermore, costs were avoided by optimisation of WWTPs, tariffs and finally by replacement of fertilisers. Avoided costs by taxes for water use and taxes for wastewater discharges directly to water bodies were negligible, since these unit costs are extremely low in Colombia. Applying realistic levies for consumer charges and 'polluter pays principles' would have a significant effect in favour of the 3-SSA.

The research described in this thesis also highlights the following outcomes:

1) A technology selection model of prevention and minimization strategies at the household level, considering different combinations of alternatives: low consumption devices, greywater use, and rainwater harvesting. In this model, alternatives were hierarchized using an analytic hierarchy process (AHP) and grey relational analysis (GRA). A cost-benefit analysis was carried out to compare the highest ranked alternatives with the conventional approach, which considered a 'business-as-usual scenario' of high water use, end-of-pipe wastewater treatment plants and the conventional water supply system with drinking water quality for all uses. For the study area, the minimization and prevention alternatives were viable ($NPV_{Benefit}/NPV_{cost}$ >1.0) when they were implemented in more than 20% of households.

2) A conceptual framework (CF) to technology selection in the urban drainage system, including strategies at the basin level, sustainable urban drainage systems (SUDS) and the integral vision of the 'sewage-WWTP-receiving water body' system. The CF was applied to the collection and transport of runoff and wastewater. The CF flow chart was designed to help decision makers in the selection of urban drainage strategies with the purpose of optimizing the investments considering cleaner production concepts.

3) A methodology to evaluate the potential of wastewater reuse in agricultural irrigation. The results showed that there are two key factors that influence the reuse potential of treated wastewater for sugarcane crop irrigation: i) the rainfall temporal variation, which defines the magnitude and time period of irrigation requirements, and ii) the costs incurred to achieve the required effluent quality.

4) Application of mathematical modelling to demonstrate the importance of considering the dynamic behaviour of the river and its pollution discharges for decision-making in water quality management. The results show that self-purification capacity can be severely affected by abrupt changes in hydraulic flows and the type and size of the received pollution from point-source and non-point source pollutants.

xii

Samenvatting

Urbanisatie en bevolkingstoename hebben verstrekkende gevolgen voor water gebruik en afvalwater management. Factoren zoals klimaatverandering maken de uitdagingen met betrekking tot waterbeheer nog belangrijker. Tegelijkertijd worden bronnen van water in sterke mate beïnvloed door menselijke activiteiten, zoals de constructie van dammen, ontbossing, erosie, mijnbouw, veranderend landgebruik en het lozen van vervuilende stoffen. In veel gevallen, in het bijzonder in ontwikkelingslanden, is de bescherming van waterbronnen tegen aantasting van de kwaliteit door punt en diffuse vervuilingsbronnen gebaseerd op 'end-of-pipe' oplossingen. Die strategie gaat alleen uit, wat betreft infrastructuur, van het bouwen van afvalwaterzuiveringen (AWZ). Vaak wordt dit vergezeld door aanpassingen in de regelgeving, inclusief het toepassen van economische instrumenten, zoals heffingen op lozingen van afvalwater. Echter, deze strategie heeft de doelstelling om de kwaliteit van waterbronnen voor verschillende toepassingen te behouden, niet volledige behaald. Doorgaan met de huidige praktijk van stedelijk waterbeheer in een 'business-as-usual' benadering is niet duurzaam, wanneer de implicaties voor de volksgezondheid, het milieu en dus de economie, in ogenschouw worden genomen. Water zou beheerd moeten worden als een eindige hulpbron die voor verschillende doeleinden gebruikt kan worden, en iedere mogelijk oplossing moet geformuleerd worden op basis van een eerlijke verdeling en criteria voor milieubescherming, uitgaande van het stroomgebied als planningseenheid. Om deze situatie te adresseren evenals de uitdagingen van de Sustainable Development Goals (SDGs) is een systematische visie noodzakelijk om de effectiviteit van investeringen in water en sanitatie te garanderen. De 2030 Agenda erkent de centrale rol van waterbronnen voor duurzame ontwikkeling en de essentiële rol die beter drinkwater, sanitatie en hygiëne speelt in de voortgang op andere gebieden, inclusief gezondheid, onderwijs en de vermindering van armoede.

Het onderzoek zoals beschreven in dit proefschrift heeft als doel bij te dragen aan de oplossing van het hierboven beschreven probleem. In dit onderzoek is een speciale methode onderzocht van technologie-selectie voor het beperken van vervuiling door huishoudelijk afvalwater. De methode van technologie-selectie bestond o.a. uit een multi-criteria analyse (MCA), de toepassing van wiskundige modellen van waterkwaliteit in rivieren en een kosten-baten analyse (KBA). Het stroomgebied is gebruikt als eenheid van analyse voor een integrale benadering met inachtneming van aspecten van technische, milieukundige, sociale, culturele, economische, beleidsmatige en regelgeving betreffende aard. Dit onderzoek was gericht op het valideren van een strategie van technologie selectie gebaseerd op het concept van de Drie-Staps Strategische Methode (3-SSM). In die context wordt 'technologie-selectie' niet gezien als zijnde alleen de waterzuiveringstechnologie, maar inclusief factoren zoals minimalisatie en voorkoming van vervuiling, zowel in de stedelijke water cyclus (huisvesting en het stedelijk drainage systeem) als op het niveau van het stroomgebied, AWZ, hergebruik van effluent en de natuurlijke en/of verbeterde zelf-zuiverende capaciteit van oppervlaktewater.

In dit promotie onderzoek is elke stap van de 3-SSM apart onderzocht. De resultaten en conclusies van het onderzoek van iedere stap afzonderlijk zijn gebruikt als startpunt voor een omvattende analyse van een achtereenvolgende implementatie van de drie stappen in chronologische volgorde. De 3-SSM (Niet conventionele strategie) is gevalideerd door toepassing op de bovenstroom van de Cauca rivier (voor het gedeelte La Balsa-Anacaro; 389 km) in Colombia. Het onderzoek omvatte ook een vergelijking met een Conventionele Strategie, die uitging van een 'business-as-usual' scenario met een hoog waterverbruik, 'end-of-pipe' afvalwaterzuivering en een conventioneel systeem voor drinkwatervoorziening met dezelfde kwaliteit voor alle verschillende soorten gebruik. De toepassing van de 3-SSM hield een vermindering van waterverbruik in (Stap 1) en hergebruik van gezuiverd afvalwater in huishoudens en voor het irrigeren van suikerriet (Stap 2). De methode bestond ook uit een prioritering van investeringen om het effect te maximaliseren voor het verbeteren van de waterkwaliteit van de Cauca rivier, door het aanpakken van bronnen van vervuiling in deelstroomgebieden en gemeenten met de grootste vervuiling die zich stroomopwaarts bevinden van segmenten van de rivier met de laagste zuurstof concentraties (Stap 3). De KBA liet duidelijk zien dat de 3-SSM (Niet conventionele strategie) betere resultaten behaalde. De 3-SSM veroorzaakte een belangrijke verbetering van de waterkwaliteit van de Cauca rivier. Dat werd vooral veroorzaakt door het grote verschil in initiële investeringen en kosten voor in bedrijf houden en onderhoud van AWZs in gemeenten voor de twee strategieën. Voor de Niet conventionele Strategie waren de AWZs kleiner door de toepassing van preventie en minimalisatie en door afvalwater hergebruik. De toepassing van de 3-SSM resulteerde in vermeden kosten voor initiële investeringen en voor de bedrijfsvoering en het onderhoud, in het bijzonder voor grondwater putten en de benodigde pompen voor irrigatie van suikerriet. Bovendien werden kosten vermeden door optimalisatie van AWZs, tarieven en tenslotte ook door het vervangen van kunstmest. Vermindering van kosten door vermindering van heffingen voor watergebruik en heffingen voor afvalwater lozingen direct in het oppervlaktewater waren verwaarloosbaar, doordat deze heffingen al extreem laag zijn in Colombia. Het opleggen van hogere heffingen voor water gebruik en voor het 'de vervuiler betaald' principe zou een significant effect opleveren ten gunste van de 3-SSA.

Het onderzoek zoals beschreven in dit proefschrift heeft tot de volgende belangrijke resultaten geleid:

1) De ontwikkeling van een technologie-selectie model gebaseerd op preventie en minimaliseringsstrategieën op het niveau van huishoudens, met gebruik van verschillende combinaties van maatregelen: water besparende apparatuur, hergebruik van grijswater en de opvang van regenwater. Dit model bepaalde ook een ranglijst van opties door middel van het analytical hierarchy process (AHP) en grey relational analysis (GRA). Een kosten-baten analyse is ook uitgevoerd om de beste opties van de ranglijst te vergelijken met de conventionele methode, die gebaseerd was op een 'business-as-usual scenario' van hoog waterverbruik, end-of-pipe afvalwaterzuivering en een conventioneel drinkwater systeem met het gebruik van drinkwater voor alle doeleinden. De optie gebaseerd op minimalisatie en

preventie was haalbaar in het bestudeerde onderzoeksgebied (NPV$_{Baten}$/NPV$_{Kosten}$ >1.0) onder voorwaarde dat de implementatie in meer dan 20% van de huishoudens werd gerealiseerd.

2) Een conceptueel kader (CK) for technologie-selectie voor het stedelijk drainage en rioolsysteem, inclusief strategieën op het niveau van het stroomgebied, duurzame stedelijke drainage en rioolsystemen (SUDS) en een integrale visie op het 'riolering-AWZ-ontvangende opervlaktewater' systeeem. Dit CK is toegepast op de inzameling en het transport van hemelwater en afvalwater. Het CK stroomschema is ontworpen om beleidsmakers te helpen met de selectie van stedelijke afwateringstrategieën met als doel de optimalisatie van investeringen voor de toepassing van 'milievriendelijke productie'.

3) Een methode om het potentieel van afvalwater hergebruik in de landbouw voor irrigatie te evalueren. De resultaten lieten zien dat er twee bepalende factoren zijn die het potentieel voor hergebruik van behandeld afvalwater voor irrigatie van suikerriet bepalen: i) de variatie in de tijd van regenval, die de behoefte bepaalt wat betreft hoeveelheid en timing van irrigatie, en ii) de kosten die gemaakt moeten worden om de gewenste effluent kwaliteit te bereiken.

4) Toepassing van wiskundige modellen om het belang te demonstreren van het meenemen van het dynamische gedrag van een rivier en de lozingen van vervuilende stoffen in de besluitvorming voor waterkwaliteitsbeheer. De resultaten laten zien dat de zelfzuiverende capaciteit negatief beïnvloed kan worden door abrupte veranderingen in het debiet in de rivier en het soort en de grootte van de lozing van vervuilende stoffen vanuit zowel puntbronnen als diffuse bronnen.

Table of contents

List of Figures

List of Tables

List of abbreviations and acronyms

3-SSA	Three Step Strategic Approach
AD	After Christ
AHP	Analytic hierarchy process
ASCE	American Society of Civil Engineers
AWWT	Anaerobic Wastewater Treatment
B/C	Benefit cost ratio
BC	Before Christ
BOD	Biochemical Oxygen demand
BOD$_5$	Biochemical Oxygen Demand - five days
CAEDYM	Computational Aquatic Ecosystem Dynamic Model
CBA	Cost-Benefit Analysis
CEPIS	Centro Panamericano de Ingeniería Sanitaria y Ciencias del Ambiente
CF	Conceptual framework
CI	Consistency index
CIRIA	Construction Industry Research and Information Association
COD	Chemical Oxygen Demand
COLCIENCIAS	Departamento Administrativo de Ciencia, Tecnologia e Innnovación
CP	Cleaner Production
CPI	Consumer Price Index
CR	Consistency ratio
CRC	Corporación Autónoma Regional del Cauca
CSO	Combined Sewer Overflows
CVC	Corporación Autónoma Regional del Valle del Cauca
DAF	Dissolved Air Flotation Unit
DANE	Departamento Administrativo Nacional de Estadística
DG	Directorate General
DO	Dissolved Oxygen
DWTP	Drinking water treatment plant
ELCOM	Enterprise Web Content management
EMCALI	Empresas Municipales de Cali
EPANET	Water Distribution System Modelling
EPSA	Empresa de Energía del Pacífico
FAL	Fotografías Aereas (aerofotogrametry company)
FAO	Food and Agriculture Organization of the United Nations
FM	Maximum Factor
GBA	Global Business Alliance
GHG	Green House Gas
GNI	Gross National Income
GOD	Groundwater occurrence, Overall aquifer class and Depth of water table
GPA	Global Programme Action

GRA	Grey Relational Analysis
GWP	Global Water Partnership
HCES	Household-centred Environmental Sanitation
HO	Helminth eggs
ICA	Instrumentation Control and Automation
ICT	Information and Communications Technology
IDB	Inter-American Development Bank
IDEAM	Instituto de Hidrología, Meteorología y Estudios Ambientales
IHE	Institute for Water Education
IHP	International Hydrological Programme
INBO	International Network of Basin Organization
IUWCM	Integrated Urban Water Cycle Planning and Management
IWM	Integrated Urban Water Management
IWRM	Integrated Water Resources Management
K	Potassium
MADS	Ministerio de Ambiente y Desarrollo Sustentable
MAUT	Multi Utility Technique
MAVDT	Ministerio de Ambiente, Vivienda y Desarrollo Territorial
MCA	Multi-Criteria Analysis
MCDA	Multi-Criteria Decision Analysis
MDG	Millennium Development Goals
MENA	Middle East and North Africa
MP	Micropollutant
MW	Megawatt
N	Nitrogen
NPV	Net Present Value
O&M	Operation and Maintenance
OD	Oxygen Demand
OECD	Organization for Economic Cooperation and Development
OPS	Organización Panamericana de la Salud
P	Phosphorus
PC	Personal Computer
POP	Persistent Organic Pollutant
PROSAB	Support system for the selection of after treatment alternatives for anaerobic reactor effluents
PROSEL	Process Selection Model
Qm	Average Flow
QUAL 2K	Water Quality modelling developed by U.S. EPA
Redox	Oxidation/reduction Potential
RI	Random average index
SAMTAC	South American Technical Advisory Committee - CEPAL

SANEX	A decision support system for selecting sanitation systems in developing countries
SDG	Sustainable Development Goals
SDR	Social Discount Rate
SDS	South Drainage System
SUDS	Sustainable Urban Drainage Systems
SUI	Public Services Unified Information System of the Republic of Colombia
SWITCH	Sustainable Urban Water Management Improves Tomorrow's City's Health (a EU supported project on 'Water in the city of the future')
SWMM	Storm Water Management Model
TMDL	Total daily maximum load
TSS	Total Suspended Solids
U.S.	United States
U.S. EPA	United States Environmental Protection Agency
UASB	Up-flow Anaerobic Sludge Blanket
UN	United Nations
UNDP	United Nations Development Programme
UNESCO	United Nations Educational, Scientific and Cultural Organization
UNESCO-IHE	Institute for Water Education
UNICEF	United Nations Children's Fund
Univalle	Universidad del Valle
USA	United States of America
UWC	Urban Water Cycle
UWM	Urban Water Management
UWOT	Urban Water Optioneering Tool
WASP	Water Quality Simulation Programme
WAWTTAR	Water and Wastewater Treatment Technologies Appropriate for Reuse
WCED	United Nations World Commission on Environment and Development
WFD	Water Framework Directive
WHO	World Health Organization
WISE	World Innovation Summit for Education
WSP	Water and Sanitation Program World Bank
WSUD	Water-Sensitive Urban Design
WWAP	World Water Assessment Programme
WWT	Wastewater Treatment
WWTP	Wastewater Treatment Plant
WWTP-C	Wastewater Treatment Plant - Cañaveralejo, Cali

Chapter 1
General introduction

Source: CVC photo file

1.1 Background and problem outline

1.1.1 Increased demand for water resources

Water is the most abundant chemical component in the biosphere and probably the most important one. Almost all living creatures, including human beings, use water as their basic means for metabolic performance. The removal and dilution of many human and natural discharges are obtained through the use of water. Likewise, water has physical characteristics that have a direct impact on the evolution of our environment and the life that has developed in it. The exponential growth rate of the human population, as well as agricultural and industrial expansion, have generated an increase in water supply demand, and consequent challenges of access. This situation has been partly solved through the construction of dams, reservoirs, changes in water streams, pipelines and aqueducts to bring water from remote, uncontaminated sources. Groundwater reportedly provides drinking water to at least 50% of the global population and accounts for 43% of all the water used for irrigation (FAO, 2010). Worldwide, 2.5 billion people rely exclusively on groundwater resources to meet their daily basic water needs (UNESCO, 2012). Groundwater supplies are diminishing, with an estimated 20% of the world's aquifers being over-exploited (Gleeson *et al.*, 2012), leading to serious consequences such as land subsidence and saltwater intrusion in coastal areas (WWAP, 2015). Additionally, technological development has resulted in satisfying the water requirements for municipal, agricultural and industrial use, increasing the competition for easy access to clean freshwater sources (Marsalek *et al.*, 2008a). Currently, 54% of the world's population lives in urban areas. By 2050, over 70% of the global population will be urban residents (United Nations, 2014). In 2015, cities accounted for 60% of global drinking water consumption, 75% of global energy consumption, and 80% of global greenhouse gas (GHG) emissions (Crittenden, 2015). The trend towards more urbanized societies and the growing number of people living in large cities has huge implications for freshwater use and wastewater management (United Nations, 2014). Currently the world is in crisis due to the quantity of available water (water scarcity). The essence of global water scarcity is the geographic and temporal mismatch between freshwater demand and availability. The increasing world population, improving living standards, over-abstraction, changing consumption patterns, and expansion of irrigated agriculture are the main driving forces for the rising global demand for water climate change, such as altered weather patterns (including droughts). A paradigm shift is urgently needed to achieve sustainable use of water resources.

1.1.2 Water pollution

In most cases, used water is returned to water resources as untreated wastewater, leading to water quality deterioration, which negatively impacts on aquatic habitats and the quality of life of communities, with subsequent economic, social and environmental impacts (Marsalek *et al.*, 2008a). Approximately 80% of wastewater is released into the environment without adequate treatment (United Nations-Water 2015). Unlike point source pollution, which enters a river course at a specific site such as a pipe discharge, diffuse pollution occurs when polluting substances leach into surface waters and groundwater as a result of rainfall, soil infiltration or surface runoff (Bravo-Inclán *et al.* 2013). More than 600 chemicals pollutants have been identified in stormwater. These chemicals can affect human health and aquatic life. The list of contaminants associated with diffuse pollution includes: solids, chloride, nutrients (N and P), pesticides, polycyclic aromatic hydrocarbons, pathogens, heavy metals, etc. (Marsalek *et al.*, 2008b). Typical examples of diffuse pollution include the use of fertiliser in agriculture and forestry, pesticides from a wide range of agricultural land uses, contaminants from roads and paved areas, and atmospheric deposition of contaminants arising from industry (Environment Agency, 2007).

In addition to the classical parameters of contamination by organic matter and pathogens associated with point-source pollution by domestic wastewater, in the last decades, there has been concern about the contamination of water by micro-pollutants (MPs). Higher concentrations of persistent organic pollutants (POPs) have been found in food chains exposing humans and wildlife to toxic effects and diseases (Fürhacker *et al.*, 2016). Thousands of chemicals play an important role in our daily activities. As a result of widespread use and poor management, these substances also enter the environment. A significant pathway for the input and spread of chemicals is water - for example, when substances are washed out by rainwater or transported by wastewater (Wittmer and Burkhardt, 2009). Grey water, which originates from the kitchen, bathroom or laundry, can contain over 900 synthetic organic compounds or xenobiotics (Erikson, 2002). Current assessments of water scarcity primarily focus on water quantity. But as water quality issues are prevalent worldwide, we need to rethink the concept of water scarcity to include also the quality of freshwater resources available for different water use sectors and ecosystems. Deforestation and water pollution are the main driving forces of this water quantity crisis. Therefore, the paradigm shift in water management must involve both water quantity and water quality.

1.1.3 Climate Change

Climate change makes the problems related to freshwater more critical. Climate change is associated with the sea level rise and the intensification of the hydrological cycle, producing more frequent and intense rainfall as well as extended dry periods. As a result, a city's water supply, wastewater and stormwater systems will be particularly affected. Climate change's impacts on the urban water system typically has knock-on effects on other urban systems because of the role that water plays in the system performance as well as in the maintenance of the quality of life in a wider sense (Novotny, 2008). A linkage between the global warming observed in recent decades and the large-scale changes in the hydrological cycle has been observed. These disturbances include changes in vapour content in the atmosphere, precipitation patterns, rainfall intensity and frequency of extraordinary storms, snowpack depth, glacier cover, soil moisture, and runoff processes (Bates *et al.*, 2008). In many lakes and reservoirs of the world, climate-change effects are mainly due to variations in water temperature affecting oxygen regimes, oxidation/reduction reactions, stratification, mixing rates, and the development of biota (Montes-Rojas *et al.*, 2015). For example, increasing the temperature decreases the self-purification capacity of rivers by reducing the amount of dissolved oxygen, which is used for biodegradation. An increase in heavy precipitation leads to increased nutrients, pathogens, and toxins in water bodies (Kundzewicz *et al.*, 2007).

1.1.4 Effect of human activities on water resources

Water resources are substantially affected by human activities such as dam building, deforestation, mining activities, land use changes and pollutant loads. Human activities can exacerbate the negative impacts of climate change by increasing the vulnerability of systems to a changing climate (Bates *et al.*, 2008). Other impacts are associated with the building of housing in sensitive areas, such as on high slopes in the upper parts of water catchment areas, and very close to sensitive groundwater aquifers. The damage to freshwater resources coincides with the increased demand for water. The erosion associated with deforestation has altered the water cycle and has caused the loss of soil, increasing the sediment load transported towards the coasts.

1.1.5 Global agendas and approaches for water resources management

Over the last few decades, a number of new concepts and terminologies related to sustainable water management have emerged. Among them, the following stand out: Resilience, Integrated Water Resources Management (IWRM), Hydrological cycle, Urban Water Cycle (UWC), Integrated Urban Water Management (IWM), Ecohydrology, and Water Governance. The validity of using water basin space, or interconnected basins, as a basic area for integrated water management was included in the recommendations resulting from the United Nations Water Conference held in Mar del Plata, Argentina, in 1977. The importance of sustainable use and provision of water was endorsed by Agenda 21 in Rio in 1992. The World Water Vision, published in the year 2000, expressed serious concerns about water availability for future generations. The World Summit on Sustainable Development in Johannesburg in 2001 identified a set of priorities in eight ambitious goals with concrete targets as formulated in the Millennium Development Goals (MDG). One goal (Goal 7) specifically addressed the challenges of access to safe water and sanitation.

1.1.6 Water quality management

The protection of water resources from quality deterioration by point and non-point source pollution discharges is probably the biggest challenge in sustainable water resources management over the decades. In the 1960s and 1970s we started to see the first signs of the 'pandemic' of water pollution. In response to this, most countries adopted pollution controls which were based on 'end-of-pipe' solutions, which considers only wastewater treatment plants (WWTPs), in terms of infrastructure, accompanied by adjustments to the regulations, including the application of economic instruments, such as taxation or penalties for wastewater discharges

(Gijzen, 2006). However, this strategy has not fully fulfilled the objective of recovering the quality of the water resource for its different uses. In many cases this approach has failed as treatment systems often operate with low efficiency or have been completely abandoned as a result of the lack of both prioritization of investments and O&M. Besides, end-of-pipe WWT does not contribute much to the control of diffuse pollution. Continuing the urban water practice in a 'business-as-usual' manner is unsustainable. This practice has resulted in significant problems related to public health, the environment and the economy.

1.1.7 Water management and Sustainable Development Goals (SDGs)

In 2015, 6.6 billion people, or 91% of the global population, used an improved drinking water source. However, in 2015 an estimated 663 million people were still using unimproved sources or surface water. Between 2000 and 2015, the proportion of the global population using improved sanitation increased from 59% to 68%. However, 2.4 billion were left behind. Among them were 946 million people without any facilities at all who continue to practise open defecation (United Nations, 2016). The following illustrate some of the serious threats to water-related sustainable development: 1) 2.1 billion people lack access to safely managed drinking water services (WHO and UNICEF, 2017); 2) 80% of wastewater effluents flows back into the ecosystem without being treated or reused (UNESCO, 2017); 3) The increased use of fertilizer for food production, combined with increased wastewater effluent, results in a 10-20% increase in nitrogen flow into global rivers (UNEP, 2007); 4) 1.8 billion people use a source of drinking water with faecal contamination and 340,000 children under five die every year from diarrhoeal diseases (UNICEF and WHO, 2015).

From MDGs to SDGs
In September 2015 the United Nations General Assembly unanimously adopted the Sustainable Development Goals (SDGs). The importance of water as an integral part of all human development and ecosystem needs is emphasized through the dedicated Water Goal SDG 6. While many of the Millennium Development Goal (MDG) targets for 2015 have been met or even passed, the MDG target of halving the share of the population without access to basic sanitation was missed by 9 percentage points. In absolute numbers, due to population growth, the total number of people without basic sanitation remained almost the same. While major resources have been allocated to health care, education and other development priorities since 2000, the sanitation gap has not been prioritized. Sanitation has therefore been identified as 'the most lagging' of all the MDG targets. Furthermore, with their focus on sanitation access and their failure to address wider issues of wastewater and excreta management, the MDGs offered little incentive for investment in more sustainable systems. Thus, much of the sanitation and wastewater management development that has already taken place will require additional investment to make it both more effective and more sustainable. The universal applicability and emphasis on integrated solutions in the SDGs and the broader 2030 Agenda provide strong arguments for investing in sustainable sanitation and wastewater management. The SDGs dedicate an entire goal to water and sanitation via SDG 6 'to ensure availability and sustainable

management of water and sanitation for all', bringing greater awareness to sanitation challenges (Andersson *et al.*, 2016).

SDG 6

SDG 6 has two targets which are directly linked to sanitation and wastewater management: Target 6.2: ... '...achieve access to adequate and equitable sanitation and hygiene for all, and end open defecation, paying special attention to the needs of women and girls and those in vulnerable situations'; Target 6.3: ... '.... improve water quality by reducing pollution, eliminating dumping and minimizing release of hazardous chemicals and materials, halving the proportion of untreated wastewater, and substantially increasing recycling and safe reuse globally'. Goal 6 goes beyond drinking water, sanitation and hygiene to also address the quality and sustainability of water resources. Agenda 2030 recognizes the centrality of water resources for sustainable development and the vital role that improved drinking water, sanitation and hygiene play in progress in other areas, including health, education and poverty reduction (United Nations, 2016). Sustainable sanitation and wastewater management is influential within many of the SDGs (Table 1.1).

Sustainable sanitation (part of Goal 6) can also make cost-effective contributions to achieving a wide variety of other SDG goals and targets (Hall *et al.*, 2016). The number of targets addressed can increase with the level of ambition in sustainable sanitation and wastewater management investments. For example, at the most basic levels of ambition (ending open defecation and preventing human exposure to pathogens and toxic substances in excreta and wastewater), improving sanitation and wastewater management could relieve a large burden of infectious disease (Goal 3), particularly child mortality. A lower incidence of disease means that fewer days of education (Goal 4) and of productive work are lost (Bos *et al.*, 2004).

If systems also aim to prevent the release of untreated wastewater into natural ecosystems and to reduce the run-off of nutrients from agricultural soil caused by fertilizer application, they could improve the status of freshwater and coastal ecosystems and the services they provide (Goal 14). Recovering and reusing the valuable resources present in excreta and wastewater also contributes to resource efficiency (Goal 12), conservation of freshwater ecosystems and restoring degraded land and soils (Goal 15) (Jenkins, 2016; WHO, 2016), and can help improve food security (Goal 2). Sustainable sanitation and wastewater management value chains provide new livelihood opportunities (Goals 1 and 8). To make tomorrow's cities liveable (Goal 11) it is necessary to introduce adequate sanitation and wastewater management. Furthermore, 'equitable access' to adequate sanitation can also help to achieve non-discrimination targets under Goal 5 by increasing equal participation in school, the workforce, institutions and public life. A lack of suitable facilities effectively excludes women, girls and people with disabilities and increases the risk of gender-based violence (Andersson *et al.*, 2016). Other goals such as Goal 7 on renewables and energy efficiency will reinforce targets related to water pollution and aquatic ecosystems by reducing levels of chemical and thermal pollution (compared to a less efficient fossil energy supply system).

Table 1.1 Beyond SDG 6: Sustainable sanitation and wastewater management can help advance other SDGs and targets

Goals	Targets
1. No poverty	1.2 - poverty in all its dimensions 1.4 - access to basic services 1.5 - resilience, reduce vulnerability, extreme events
2. Zero hunger	2.1 - end hunger / food sufficiency 2.2 - end malnutrition 2.3 - double smallholders' productivity & incomes
3. Good health & well-being	3.2 - end preventable infant and under-5 deaths 3.3 - end epidemics & combat water-related diseases 3.9 - reduce deaths & illnesses from pollution and contamination
4. Quality education	4.5 - eliminate gender disparities in education 4a - build & upgrade safe education facilities
5. Gender equality	5.1 - end discrimination against women & girls 5.2 - eliminate violence against women & girls in public spaces
6. Clean water & sanitation	6.2 - sanitation & hygiene for all 6.3 - reduce water pollution, increase recycling 6.4 - substantially reduce water scarcity 6.5 - water resources management, trans-boundary cooperation 6.6 - protect & restore water-related ecosystems 6a - international cooperation, support developing countries
7. Affordable & clean energy	7.2 - increase share of renewable energy
8. Decent work & economic growth	8.4 - improve resource efficiency, decouple economic growth from environmental degradation
9. Industry, innovation & infrastructure	9.4 - upgrade industrial resource efficiency & clean technology
11. Sustainable cities & communities	11.5 - reduce deaths & econ. losses from disasters 11.6 - reduce adverse environmental impact of cities 11.7 - safe public spaces
12. Sustainable consumption & production	12.2 - management & efficient use of resources 12.4 - chemicals and waste management 12.5 - reduce waste generation
13. Climate action	13.1 - strengthen resilience to climate-related hazards & natural disasters
14. Life below water	14.1 - reduce marine pollution from land-based activities
15. Life on land	15.1 - conserve, restore & sustainably use terrestrial ecosystems 15.3 - restore degraded land and soils

Source: adapted from Andersson *et al.* (2016)

Climate change (Goal 13) will be manifested mainly by sea level rise and the intensification of the hydrological cycle, producing more frequent and intense rainfall as well as extended dry periods. As a result, a city's water supply, wastewater and stormwater systems will be particularly affected. Constructing new greener infrastructures, retrofitting or reconfiguring existing infrastructure systems and exploiting the potential of smart technologies (Goal 9) can greatly contribute to the reduction of environmental impacts and disaster risks as well as the construction of resilience and increased efficiency in the use of water resources (GWSP, 2015).

The need for a paradigm shift in urban water management
To confront these challenges, in particular the water pollution problems, it is necessary to develop new strategies that guarantee the sustainability of investments within a general framework (Sustainability, Resilience IWRM, UWM, UWC, Cleaner production, etc.), but through more defined strategies that may turn into concrete actions. In order for investments in water and sanitation to produce the expected outcomes in quality of life improvement in communities, a holistic vision of the problem is necessary. Water should be administered as a limited resource with multiple uses, and solutions should be formulated with appropriate distribution (for its different uses) and protection criteria, considering the basin as a planning unit.

The paradigm associated with the sustainable city of the future should include strategies such as: 1) *New generation systems* (prevention and minimization, the nutrient cycle: closing the loop, sustainable urban drainage systems SUDS, natural systems for treatment, urban agriculture, etc.); 2) *'Run to failure',* stop repairing the old systems and gradually replace them with new generation systems (Nelson, 2008); 3) *Decentralization; 4) Instrumentation, Control and Automation (ICA).* A city of the future may also be defined as an 'EcoCity'. An 'EcoCity' is a city that balances social, economic and environmental factors to achieve sustainable development. This concept is also captured under different terminologies such as Green Cities, 'Blue-Green cities', and 'Water Sensitive Urban Design'. *'Green cities',* in which the green infrastructure is implemented, consist of those parts that contribute to the natural processes of keeping the water and the recycling of waste (Fletcher *et al.*, 2015). *'Blue-Green cities'* aim to recreate a naturally-oriented water cycle while contributing to the amenity of the city by bringing water management and the green infrastructure together (Everett *et al.*, 2015*). Water Sensitive Urban Design (WSUD)* is based on the integration of two key fields including 'Integrated urban water cycle planning and management' (IUWCM) and 'urban design'. For this it is necessary to integrate urban design with the various disciplines of engineering and environmental science associated with the provision of services and the protection of aquatic environments in urban areas (Wong and Ashley, 2006). These concepts will be explained in more detail in Chapter 2.

1.1.8 Scope of this PhD research

The research described in this thesis intends to contribute to realizing a paradigm shift towards sustainable water management in the city of the future. The research is aimed at identifying and validating a suitable and innovative comprehensive strategy for sustainable urban water management based on cleaner production principles. The methodology involved multi-criteria analysis (MCA) and cost-benefit analysis (CBA). The methodology integrated technical, environmental, social, cultural, economic, policy and regulatory aspects. This research was oriented towards the development of a strategy of technology selection based on the 3-Step Strategic Approach concept (3-SSA) (Nhapi and Gijzen, 2005; Gijzen, 2006), not only in the urban water cycle, but also in the basin, considering it as the unit of analysis. The development of the strategy included, among other factors, the priority water uses and the wastewater pollution control plans for both the medium and long term. In this context, 'technology selection' shall not be understood as the treatment technology but will include aspects such as minimisation and prevention, both in the urban water cycle (housing and urban drainage system) and at the basin level, WWTPs, reuse of effluents, and the natural and/or stimulated self-purification capacity of the water bodies. The study area for the development was the Upper Cauca river basin of the Cauca River, the second most important river in Colombia. The research included a comparison between the conventional strategy (end-of-pipe solution) and unconventional strategies (3-SSA).

1.2 Research objectives

1.2.1 Overall aim

To identify and validate a suitable and innovative strategy for sustainable urban water management based on cleaner production principles, which considers prevention/minimization, treatment for reuse, and stimulation of water resource self-purification capability as part of a comprehensive approach.

1.2.2 Specific objectives

i) To identify and validate ways to maximize prevention/minimization of pollution by various options and interventions in the municipal water cycle;

ii) To identify and validate ways to maximize treatment for recovery and reuse in the municipal water cycle;

iii) To identify and validate ways to maximize the self-purification potential of water bodies considering the river basin as the unit of analysis;

iv) To identify and validate ways to maximize the combined impact of interventions, relating to prevention/minimization, treatment for reuse, and self-purification considering the basin-like unit of analysis.

1.3 Study area: Upper Cauca river basin, Colombia

1.3.1 General characteristics

The Cauca River is the second most important fluvial artery of Colombia and the main water source of the Colombian southwest. It has a length of 1,204 km with a basin of 59,074 km², which represents 5% of the territory of Colombia (Sandoval and Ramírez, 2007). Along this river basin there are 183 municipalities and about 12.5 million people, which represents approximately 25% of the Colombian population. The Cauca's river basin is divided and classified in three sections (Figure 1.1)

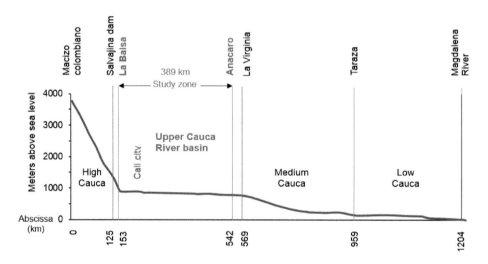

Figure 1.1 Profile of Cauca River
Source: Adapted from Sandoval and Ramirez (2007)

The Cauca River's geographical valley (the Salvajina dam to the Anacaro Station) is considered one of the most fertile areas in Colombia, and is the base for an important part of the Colombian economy. It has a tropical climate. The average temperature ranges between 21°C and 24°C. The duration of sunlight is longer during the dry months and shorter during the wet season. The average monthly humidity is between 70% and 75%, and the evaporation is between 1,100 mm and 1,300 mm. The average annual rainfall varies between 938 mm and 1,882 mm. The study area is the Upper Cauca river basin (Figure 1.2). It is located between the Balsa and Anacaro stations, with a length of 389 km (32% of the total length of the Cauca River) and has about 25% of the total area of the Cauca river basin. This stretch of the river has an average width of 105 m. The depth can vary between 3.5 and 8.0 m. The longitudinal profile of the river shows a concave shape and a hydraulic slope, which varies between 1.5×10^{-4} m/m and 7×10^{-4} m/m (Ramirez *et al.*, 2010). The sugar cane crops and the Colombian sugar industry are located in the flat area along the Upper Cauca river basin. In the mountain area, there are coffee crops and associated industry. There are also other farming developments, and other economic activities such as mining and manufacturing (CVC, 2015).

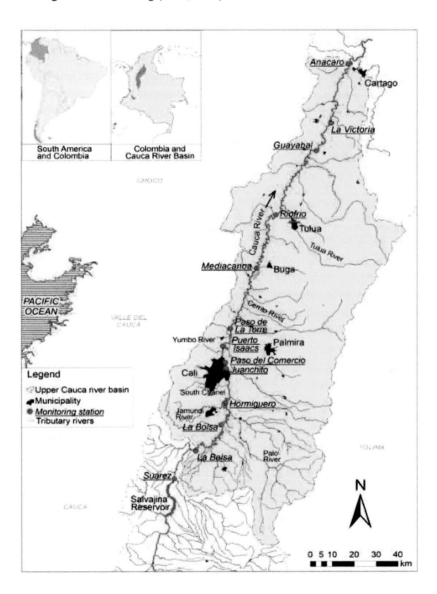

Figure 1.2 General Location of the Upper Cauca river basin, the La Balsa-Anacaro stretch
Source: Adapted from Sandoval and Ramirez (2007)

1.3.2 Water uses and water quality

The Cauca River has been used for fishing, recreation, energy generation, riverbed matter extraction, human consumption, irrigation and industry. The Salvajina reservoir started operation in 1985 and is part of the flow regulation project of the Cauca River, implemented for flood control, improvement of water quality and power generation (Galvis, 1988). The reservoir power station has a capacity of 270 MW. The reservoir operates between levels of 1,110 and 1,150 meters above sea level (m.a.s.l.), it has a minimum discharge of 60 m³/s and an average daily flow rate of 140 m³/s in the Juanchito Station (Sandoval *et al.*, 2007).

The Cauca River is also used as a receiving water body for solid waste and dumping of industrial and domestic wastewater, which is contributing to the deterioration in water quality (Figure 1.3). In the study area, there are currently 3.8 million inhabitants who form the source for approximately 134 T/d of BOD_5 to the Cauca River in the study reach. In addition to organic matter (measured in terms of BOD_5), the river has other types of associated contaminants with acute risk (coliforms and turbidity) and chronic risk (colour, phenols, heavy metals, pesticides and emerging pollutants). For the stretch of the Cauca River considered in this study, self-purification capacity was heavily affected by abrupt changes in its dilution ability and by the type, size and spatial distribution of the received pollution. For the Cauca River, most of the self-purification capacity has been lost in the last 60 years. For instance, a wetland area of 300 km^2 in the 1950s was reduced in 1986 by 90% (Muñoz, 2012). In addition, the most important wetland, the Sonso lagoon, reduced its surface area from 623 ha in 1989 to 230 ha in 2009 (Figueroa-Casas, 2012).

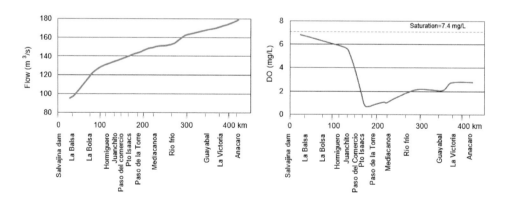

Figure 1.3 Cauca River. Typical flow and DO, the La Balsa-Anacaro stretch, dry season 2013
Based on: (CVC, 2015)

1.3.3 Strategies for pollution control

Efforts to improve the water quality of the Cauca River in the study area (the La Balsa-Anacaro stretch) began more than 50 years ago. Most of the actions have focused on the control of point-source pollution. Initially the actions were focused on stimulating the construction of WWTP in the industrial sector, while in recent decades emphasis has also given to the construction of municipal and domestic WWTPs. Economic instruments have also been applied to control pollution, such as taxation for effluent discharges, both for the industrial and the domestic sectors. Of the 38 municipalities in the study area, only 19 municipalities (50%) have a WWTP. However, these strategies have not yielded the expected results and the water quality of the Cauca River has progressively deteriorated, despite large investments in the control of contamination. Figure 1.4 shows how the BOD₅ discharged into the Cauca River and the minimum DO have changed in the La Balsa-Anacaro stretch during the period 1963-2014. Another indicator of the poor water quality of the Cauca River are the quality indexes reported by the environmental authority (CVC) and the frequent closures of the intake of water supply of Cali city, that supplies 2 million people. The frequency of closures, associated with the poor quality of the Cauca River, went up from 10 in the year 2000 to 43 in 2016 (Almario and Duque, 2017). These closures have a duration from a few hours up to two days. Pollution peaks are coming from different sources upstream of Cali and they are associated with diffuse pollution sources such as runoff from rural and urban areas. Additionally, the re-suspension of sediments and solid waste accumulated in the drainage network following heavy rainfall is also contributing to the occurrence of pollution peaks (Moreno, 2014).

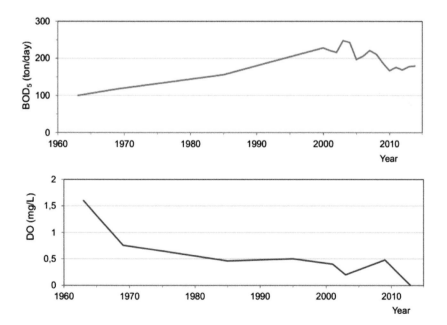

Figure 1.4 BOD₅ discharged into the Cauca river and minimum DO in the La Balsa-Anacaro stretch.
Typical dry season condition during the period 1963-2014
Sources: Donaldson et al., (1963); CVC (1971, 1976, 2004, 2010, 2015); Emcali (1985); Univalle and
CVC (2004, 2007, 2009); Univalle and Emcali, (2006)

Pollution control policies for water quality improvement of the Cauca River have not been successful. This is partly explained by the increase in pollutant loads, several WWTP being out of operation, and those WWTP that are working having operation and maintenance problems. Besides, the implemented strategies have not considered the impact of diffuse pollution associated with the agricultural sector, surface runoff at urban and rural levels and the inadequate management of solid waste. This type of pollution has been considered in the local and national public policy documents, but no concrete action has been taken to control this type of pollution.

To face this situation, it is necessary to address the structural problems of the deterioration of water quality in the Cauca River. It is necessary to strengthen inter-institutional work, citizen participation, efforts to establish water quality monitoring, and to implement innovative strategies and technologies, encouraging the use of cleaner production practices in the domestic and industrial wastewater management. The results of this research can contribute to the formulation of a vision shared by different stakeholders related to water quality improvement of the Cauca River. Considering the basin as the unit of analysis and with good leadership, it could be possible for these stakeholders to agree on a coordinated work plan and to define the short, medium and long-term activities and the prioritization of investments to achieve the vision.

1.4 Outline of the Thesis

The general structure of the thesis is presented in Figure 1.5. The document contains 8 chapters. A description of the content of each chapter is presented in what follows.

Chapter 1. General Introduction. Chapter 1 describes the problem under study, the relevance of the research topic, the research objectives and a brief description of the study area.

Chapter 2. This chapter corresponds to the literature review. It starts with the historical background of water pollution control and the origin, evolution and crisis of the conventional strategy, focused on the implementation of WWTPs (end-of-pipe solutions). In this chapter some concepts of sustainable water management are reviewed, from which innovative strategies are constructed. A short description of one of these strategies, the Three-Step Strategic Approach (3-SSA), is presented in Chapter 2. This research was based on the validation of this strategy. The final part of the chapter presents some considerations about water management in the city of the future and water resources management in the context of the Sustainable Development Goals.

Chapter 3. This chapter is focused on the first step of the 3-SSA: prevention and minimization, considering different alternatives to reduce water use at the household level. These alternatives include: change of habits, low consumption devices, rainwater harvesting, and grey water reuse. The alternatives were hierarchized using an analytic hierarchy process and grey relational analysis. A cost-benefit analysis was carried out to compare the highest ranked alternatives with the conventional approach, which considers a 'business as usual scenario' of high water use, end-of-pipe wastewater treatment and the conventional water supply system with drinking water quality for all uses. The assessment includes a case study in the expansion area of the city of Cali, Colombia.

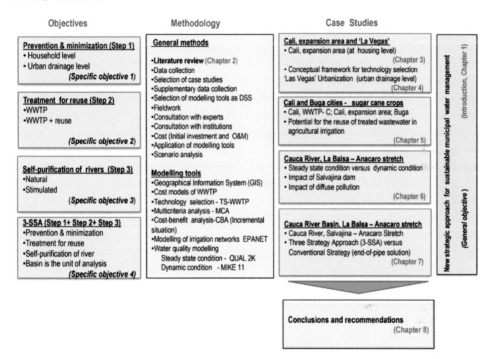

Figure 1.5 General structure of the thesis

Chapter 4. This chapter is also focused on the first step of 3-SSA: prevention and minimization. This chapter presents the development and application of a conceptual framework (CF) for technology selection for urban drainage. The CF is applied to the collection and transport of runoff and wastewater, but does not include technology selection for WWTP. The CF can be applied in new urban areas and in the expansion areas of existing cities. The CF was based on the 3-SSA and it was developed for urban conditions in the cities of the Upper Cauca river basin (Colombia). The flow chart of the CF was designed to help decision makers in the selection of urban drainage strategies with the purpose of optimizing the investments that consider cleaner production concepts.

Chapter 5. This chapter presents the potential of reuse of treated wastewater for the irrigation of sugarcane crops in the Upper Cauca river basin in Colombia. This research corresponds to the second step of the 3-SSA. The study included three case studies, with different characteristics of wastewater, flows, rainfall levels and irrigation requirements. A Cost-Benefit

Analysis (CBA) was used to compare the options: with and without reuse of treated wastewater. The results of the CBA showed that there are two key factors that influence the reuse potential of treated wastewater for sugarcane crop irrigation: 1) the rainfall temporal variation, which defines the magnitude and time period of irrigation requirements, and 2) the costs incurred to achieve the required effluent quality.

Chapter 6. This chapter corresponds to Step 3 of the 3-SSA. Steady state and dynamic conditions of quantity and quality were studied in the Cauca River (the La Balsa-Anacaro Stretch). The impact of pollution from wastewater discharges under these two flow conditions was compared. A multipurpose reservoir (the Salvajina dam) was built in 1985 for pollution control (dilution capacity), power generation and flood control. The quantity and quality water models of the Cauca River were implemented in the software MIKE 11. The results of this chapter show that self-purification capacity in the Cauca River is heavily affected by abrupt changes in hydraulic flows, especially due to the operation of the Salvajina reservoir and the type and size of the received pollution from point-source and non-point source pollutants.

Chapter 7. This chapter describes the results of a study where the three steps of 3-SSA were included sequentially. This study used the results of the research presented in chapters 3, 4, 5 and 6. The full 3-SSA (non-conventional strategy) was applied in the Upper Cauca river basin (in Colombia) and it was compared with the Conventional Strategy, which considers a 'business as usual scenario' of high water use, end-of-pipe wastewater treatment and conventional water supply providing drinking water quality for all uses. In this research, the non-conventional Strategy (3-SSA) includes reduction in water consumption (Step 1), and reuse of treated wastewater in households and for sugarcane crop irrigation (Step 2). It also considers prioritization of investments to maximize the impact in improving the water quality of the Cauca River in the study area, targeting interventions in watersheds and municipalities with the highest pollutant load and located upstream of the river segments with the lowest DO (Step 3). The software MIKE 11 was used for modelling BOD_5 and DO behaviour in the Cauca River for each strategy. Additionally, the two strategies were compared using cost-benefit analysis (CBA).

Chapter 8. The conclusions and recommendations are presented in this chapter.

1.5 References

Almario, G. and Duque, A. (2017). Deterioro progresivo de la principal Fuente de abasto de la ciudad de Cali: El Río Cauca. Incidencia del Sistema de Drenaje Sur. In Workshop 'El Sistema de Drenaje Sur del Municipio de Cali. En la búsqueda de estrategias para una gestión sostenible y resiliente'. Universidad del Valle, Cali, Colombia, 26 May 2017. (In Spanish).

Andersson, K., Rosemarin, A., Lamizana, B., Kvarnström, E., McConville, J., Seidu, R., Dickin, S. and Trimmer, C. (2016). *Sanitation, Wastewater Management and*

Sustainability: from Waste Disposal to Resource Recovery. Nairobi and Stockholm: United Nations Environment Programme and Stockholm Environment Institute.

Bates, B.C., Kundzewicz, Z.W., Wu, S. and Palutikof, J.P. (Eds.) (2008). Climate change and water. Technical paper of the Intergovernmental Panel on Climate Change, IPCC Secretariat, Geneva, 210 pp.

Bos, J.J., Gijzen, H.J., Hilderink, H.B.M., Moussa, M., Niessen, L.W. and De Ruyter-van Steveninck, E.D. (2004). Quick Scan Health Benefits and Costs of Water Supply and Sanitation. Netherlands Environmental Assessment Agency (RIVM), Institute for Water Education - (UNESCO-IHE), Erasmus University Rotterdam. Bilthoven, the Netherlands.

Bravo-Inclán, L., Saldaña-Fabela, P., Dávila, J. and Carro, M. (2013). La importancia de la contaminación difusa en México y en el mundo. Technical Report. (In Spanish).

Corporación Autónoma Regional del Valle del Cauca - CVC (1971). Contaminación de corrientes Hoya del Río Cauca, Informe CVC # 71-18. Santiago de Cali, Valle del Cauca, Colombia. (In Spanish).

Corporación Autónoma Regional del Valle del Cauca - CVC (1976). Programa para el control de la contaminación del Rio Cauca. Santiago de Cali, Valle del Cauca, Colombia. (In Spanish).

Corporación Autónoma Regional del Valle del Cauca - CVC (2004). Génesis y desarrollo de una visión de progreso - 50 años CVC. Santiago de Cali, Valle del Cauca, Colombia. (In Spanish).

Corporación Autónoma Regional del Valle del Cauca - CVC (2010). Boletín Hidroclimatológico - Año 2010. (In Spanish).

Corporación Autónoma Regional del Valle del Cauca - CVC (2015). Plan de Gestión Ambiental Regional PGAR 2015-2036. Santiago de Cali, Valle del Cauca, Colombia. (In Spanish).

Crittenden, J.C. (2015). Water for Everything and the Transformative Technologies to Improve Water Sustainability. National Water Research Institute. The 2015 Clarke Prize Lecture.

Donaldson, J., Martin, L., Whitmore, F., Santamaria, L. and Paterson, J. (1963). A study of the pollution and the natural purification of the Cauca River S.A. Tulane University and Universidad del Valle. Santiago de Cali, Colombia.

Empresas Municipales de Cali-EMCALI (1985). Proyecto de tratamiento de aguas residuales de Cali. Ingesam LTDA y EMCALI. Santiago de Cali. (In Spanish).

Environment Agency (2007). The unseen threat to water quality. Diffuse water pollution in England and Wales report, May 2007.

Erickson, B.E. (2002). Analyzing the ignored environmental contaminants. *Environ Science and Technology*: 36 (7): 140–145.

Everett, G., Lamond, A. and Lawson, A. (2015). Green Infrastructure. In: Sinnett, D., Burgess, S. and Smith, N. (Eds.) Handbook on Green Infrastructure: *Planning, Design and Implementation*. Cheltenham, Bristol, UK, pp. 50-66.

FAO (2010). The Wealth of Waste: The economics of wastewater use in agriculture. Rome, FAO.

Figueroa-Casas, A. (2012). Descripción de área de estudio. In Peña, E.J., Cantera, J.R. and Muñoz, E. (Eds.) Evaluation of pollution in aquatic ecosystems. Case study in the Sonso Lagoon, Upper Cauca River Basin (pp. 53-61). Universidad Autonoma de Occidente and Universidad del Valle, Editorial Program, Cali, Colombia. (In Spanish).

Fletcher, T.D., Shuster, W., Hunt, W.F., Ashley, R., Butler, D., Arthur, S., Trowsdale, S., Barraud, S., Semadeni-Davies, A., Bertrand-Krajewski, J.L., Mikkelsen, P.S., Rivard, G., Uhl, M., Dagenais, D. and Viklander, M. (2015). SUDS, LID, BMPs, WSUD and more – The evolution and application of terminology surrounding urban drainage. *Urban Water*, 12, 1-18.

Fürhacker, M., McArdell, S.M., Lee, Y., Siegrist, H., Ternes, T.A., Li, W. and Hu, J. (2016). Assessment and Control of Hazardous Substances in Water. In Global Trends & Challenges in Water Science, Research and Management. A compendium of hot topics and features from IWA Specialist Groups Second Edition. International Water Association (IWA) Alliance House, London, UK. ISBN 9781780408378.

Galvis, A. (1988). Water quality simulation of the Cauca River. Calibration, verification and application. MSc Thesis in System and Industrial Engineering, Universidad del Valle, Cali, Colombia. (In Spanish).

Gijzen, H.J. (2006). The role of natural systems in urban water management in the City of the Future - A 3- Step Strategic Approach. *Ecohydrol. Hydrobiol. 6, 115-122.*

Gleeson, T., Wada, Y., Bierkens, M.F.P. and Van Beek, L.P.H. (2012). Water balance of global aquifers revealed by groundwater footprint. Nature, 488: 197-200, doi: 10.1038/nature11295.

GWSP (2015). Towards a Sustainable Water Future - Sustainable Development Goals: A Water Perspective. Fileadmin/images/SDG_CONF/Towards_a_Sustainable_Water_Future.

Hall, N., Richards, R., Barrington, D., Ross, H., Reid, S., Head, B., Jagals, P., Dean, A., Hussey, K., Abal, E., Ali, S., Boully, L. and Willis, J. (2016). Achieving the UN Sustainable Development Goals for water and beyond, Global Change Institute, The University of Queensland, Brisbane.

Jenkins, M. (2016). Access to Water and Sanitation. Parliamentary Office for Science and Technology (POST) note (Vol. 521). London.

Kundzewicz, Z.W., Mata, L.J., Arnell, N.W., Doll, P., Kabat, P, Jimenez, B, Miller, KA, Oki, T, Sen, Z. and Shiklomanov, I.A. (2007). Freshwater resources and their management. In: Parry, M.L., Canziani, O.F., Palutikof, J.P., Van der Linden, P.J. and Hanson, C.E. (Eds.) Climate change 2007: impacts, adaptation and vulnerability. Contribution of Working Group II to the fourth assessment report of the Intergovernmental Panel on Climate Change. Cambridge University Press, Cambridge, pp 173–210.

Marsalek, J., Jimenez-Cisneros, B., Karamouz, M., Malmquist, P., Goldenfum, J., and Chocat, B. (2008a). Urban water cycle processes and interactions. *Urban Water Series –* UNESCO - PHI. ISBN 978-0-415-45347-9.

Marsalek, J., Rousseau, D., Van der Steen, P., Bourges, S. and Francey, M. (2008b). Ecosensitive approaches to managing urban aquatic habitats and their integrations with urban infrastructure. In Wagner, I., Marsalek, J. and Pascal, B. (Eds.) Aquatic habitats in sustainable urban Water Management. Science Policy and Practice. *Urban Water Series - UNESCO – PHI.* ISBN 978-0-415-45351-6.

Montes-Rojas, R.T., Ospina-Noreña, J.E., Gay-Garcia, C., Rueda-Abad, C. and Navarro-Gonzalez, I. (2015). Water-Resource Management in México. In Setegn, S.G. and Donoso, M.C. (Eds.) Sustainability of Integrated Water Resources Management Water Governance, *Climate and Ecohydrology* (pp. 215-244). Springer International Publishing Switzerland. ISBN 978-3-319-12194-9.

Moreno, G.E. (2014). Environmental analysis of the Upper Cauca River Basin, its main actors and its impact on water supply of Cali city. What should we do? MSc thesis in Industrial Engineering, Universidad ICESI, Cali, Colombia. (In Spanish).

Muñoz, E. (2012). Wetlands in the American continent. Concepts and classification. In Peña, E.J., Cantera, J.R. and Muñoz, E. (Eds.) Evaluation of pollution in aquatic ecosystems. Case study in the Sonso Lagoon, Upper Cauca river basin (pp. 23-49). Universidad Autonoma de Occidente and Universidad del Valle, Editorial Program, Cali, Colombia. (In Spanish).

Nelson, V.I. (2008). Viewpoint: Truly Sustainable Water Infrastructure It's time to invest in next-generation decentralised technologies. *WE&T Magazine, Water Environment Federation,* September, Vol. 20, No. 9.

Nhapi, I. and Gijzen, H.J. (2005). The 3-Step Strategic Approach to sustainable wastewater management. *Water SA,* 31(1), pp. 133-140.

Novotny, E. (2008). A new paradigm of sustainable urban drainage and water management, in Oxford Roundtable Workshop on Sustainability, Paper presented at the Oxford Roundtable Workshop on Sustainability - Oxford University. pp. 1-27.

Ramírez, C.A., Santacruz, S., Bocanegra, R.A. and Sandoval, M.C. (2010). Salvajina reservoir incidence on the flow regime of the Cauca River in its upper valley. *Ingeniería de los Recursos Naturales y del Ambiente,* Vol. 9, pp. 89-99. (In Spanish).

Sandoval, M.C. and Ramírez, C. (2007). El río Cauca en su valle alto: Un aporte al conocimiento, Cali. (In Spanish).

Sandoval, M.C., Ramirez, C.A. & Santacruz, S. (2007). Monthly operating rule optimisation of Salvajina reservoir, (In Spanish). *Ingeniería de los Recursos Naturales y del Ambiente,* 6(2007), pp 93-104. (In Spanish).

UNEP (2007). Global Environmental Outlook (GEO 4) Environment for Development. Available online at:
http://www.unep.org/geo/GEO4/report/ GEO-4_Report_Full_en.pdf.

UNESCO (2012). World's groundwater resources are suffering from poor governance. UNESCO Natural Sciences Sector News. Paris, UNESCO.

UNESCO (2017). The United Nations World Water Development Report 2017. Wastewater Untapped Resource. Executive summary. United Nations World Water Assessment

Programme Office for Global Water Assessment Division of Water Sciences, UNESCO 06134 Colombella, Perugia, Italy.

UNICEF and WHO (2015). Progress on Sanitation and Drinking Water - 2015 Update and MDG Assessment (Geneva).

United Nations (2014). World Urbanization Prospects: The 2014 Revision, Highlights (ST/ESA/SER.A/366). United Nations, Department of Economic and Social Affairs, Population Division: New York, NY. 2.

United Nations –Water (2015). Wastewater Management: A UN-Water Analytical Brief. UN-Water.www.unwater.org/fileadmin/user_upload/unwater_new/docs/UN-Water_Analytical_Brief_Wastewater_Management.pdf

United Nations (2016). The Sustainable Development Goals Report, New York USA.

Universidad del Valle and Corporación Autonama Regional del Valle del Cauca CVC (2004). Sampling campaign for calibration purpose of Cauca River water quality model. Project report Volume VI, Cali, Colombia. (In Spanish).

Universidad del Valle and Corporación Autonama Regional del Valle del Cauca CVC (2007). Optimization of the water quality simulation model of the Cauca river. La Balsa – Anacaro stretch. Project report Volume XIII. Cauca river Modelling Project (PMC), Phase III. Cali, Colombia. (In Spanish).

Universidad del Valle and Corporación Autonama Regional del Valle del Cauca CVC (2009). Water quality modelling scenarios of the Cauca river. Project report, Cali, Colombia. (In Spanish).

Universidad del Valle and Empresas Publicas Municipales de Cali EMCALI (2006). Impact assessment of the proposed strategies by EMCALI for the management of wastewaters in the Cali city on the Cauca River water quality, Project Report, Cali, Colombia. (In Spanish).

WHO (2016). Protecting surface water for health. Identifying, assessing and managing drinking-water quality risks in surface-water catchments.

WHO and UNICEF (2017). Progress on Drinking Water, Sanitation and Hygiene: 2017 Update and SDG Baselines. Geneva.

Wittmer, I. and Burkhardt, M. (2009). Dynamics of biocide and pesticide input. In Anthropogenic micropollutants in water: impacts - risks - measures. *Federal Institute of Aquatic Science and Technology. Eawag News N° 67e*, pp. 4-11.

Wong, T.H.F and Ashley, R. (2006). International Working Group on Water Sensitive Urban Design, submission to the IWA/IAHR Joint Committee on Urban Drainage, March 2006.

WWAP (United Nations World Water Assessment Programme) (2015). The United Nations World Water Development Report 2015: Water for a Sustainable World. Paris, UNESCO.

Chapter 2
Literature Review

Source: CVC photo file

2.1 Historical background of water pollution control

2.1.1 The early history of water and sanitation

Some 10,000 years ago, at the onset of the Neolithic Revolution when people started adopting an agrarian way of life, humankind established permanent settlements. This new type of livelihood eventually spread everywhere and the population began to expand faster than ever before. Sedentary agricultural life made it possible to construct villages, cities and eventually states, all of which were highly dependent on water. This created a brand-new relation between humans and water. Pathogens transmitted by contaminated water became a very serious health risk for the sedentary farmers. In these urbanised environments, guaranteeing pure water for people became a prerequisite for successful urbanization and state formation. The earliest known permanent settlement that can be classified as urban is Jericho (8000-7000 BC), located near springs and other water bodies. In Egypt, there are traces of wells, and in Mesopotamia of stone rainwater channels, from 3000 BC (Marsalek *et al.*, 2008). From the early Bronze Age city of Mohenjo-Daro, located in modern Pakistan, archaeologists have found hundreds of ancient wells, water pipes and toilets. The first evidence of the purposeful construction of the water supply, bathrooms, toilets and drainage in Europe comes from Bronze Age Minoan (and Mycenaean) Crete in the second millennium BC. The experience of humankind from the very beginning testifies to the importance and safety of groundwater as a water source, particularly springs and wells. The way in which water supply and sanitation was organized was essential for early agricultural societies. If wells and toilets were in good shape, health problems and environmental risks could be avoided (Juuti *et al.*, 2008).

The history of sanitation goes back to early historic times. In the Mesopotamian Empire (3500-2500 BC), some homes were connected to a stormwater drain system. In Babylon, latrines were connected to vertical shafts in the ground. In the Indus River Valley (currently Pakistan), from about 2500 to 1500 BC, many houses had drains that led to closed sewers (Cooper, 2001). Some earthenware pipes, latrines and cesspools were connected to drainage systems in the streets. At the King Minos Royal Palace at Knossos, Crete, by 1700 BC, four separate drainage systems emptied waste through terracotta pipes (GWP and INBO, 2009). The oldest known flushing device, a latrine with a rooftop reservoir, served King Minos (2600 BC) and was reborn 3000 years later. In Greece (300 BC to 500 AD), public latrines were drained into sewers, which carried sewage and stormwater to a collection basin outside the city. Brick-lined conduits transported the wastewater to agricultural lands to irrigate and fertilize crops and fruit orchards (Burian and Edwards, 2002; Chanan *et al.*, 2013). During this period technologies and infrastructure for water surface conduction and groundwater extraction were also developed (Novotny and Brown, 2007).

Other civilizations such as the Etruscan civilization in Italy (600 BC) and the Roman civilization showed considerable progress. The Romans used rainwater collection and aqueducts extensively in their drainage system. Typically, rainwater falling onto an urban area was stored for local use. Rainfall on rooftops was often collected into a cistern located in the house (Burian and Edwards, 2002). The Roman linkage of the urban water supply and drainage systems marks one of the earliest cases of establishing an urban water cycle. Following the fall of the Roman Empire (476 AD), cities in most of Europe and parts of Asia began to shrink considerably by migration from the urban centres. This time was considered as the Dark Ages period. The population reduction of the cities resulted in the abandonment of municipal services. In the Dark Ages, sanitation practices regressed to a primitive level (GWP and INBO, 2009). The sewers implemented in Europe following the fall of the Roman Empire were simply open ditches, essentially reverting to the practices used before the Romans advanced urban drainage systems such as underground sewers. To combat the nuisance conditions arising from open channels, these were subsequently covered. Urban stormwater runoff and later industrial wastewater were the main waste discharges, which were subsequently channelled into local streams and rivers. Human faeces were collected and used in backyard gardens. Other garbage and household waste was typically stockpiled near the city or fed to pigs. The disposal of human faeces gradually became an issue in Paris and London during the Middle Ages as populations expanded in these cities. Waste disposal in Paris was unregulated until a decree in 1530 required property owners to construct cesspools in each new dwelling (Reid, 1982).

Residential wastewater management in 17th century colonial America consisted primarily of a privy with the outlet constructed at ground level, usually discharging into the yard, street, gutter or an open channel to drain the urban stormwater runoff. Because population densities were low, privies constructed in this way did not create sanitation problems, but once population increased, sanitation problems and nuisances also increased (Burian et al., 2000). The beginning of modern urban drainage practices was initiated in European cities during the nineteenth century (Reid, 1982). During the first half of the 19th century, most sewers were designed exclusively for stormwater drainage. Sanitary waste accumulated in privy vaults and cesspools and was periodically collected by scavengers and transported to a suitable disposal location. As the 19th century progressed, the concept of urban drainage changed with the incorporation of flush toilets and water-carriage sanitary waste collection into the urban drainage systems.

2.1.2 Nineteenth and twentieth centuries and 'end-of-pipe' solutions

Impact of the industrial revolution
At the beginning of the 19th century, diseases such cholera, dysentery and typhoid fever were major threats. With the arrival of the Industrial Revolution at the end of the 19th century, industries did not consider the additional load discharged into water resources as an important issue. In fact, most of the industries settled along the riverbanks for convenient access to freshwater and for waste disposal. The 'solution to pollution is dilution' belief disseminated in the entire world as countries developed more industries and a greater amount of people moved to urban areas to live and work (Cooper, 2001). The 19th century brought the emergence of large metropolises, where the increase in the number of inhabitants and urbanization processes caused rapid growth in pollution of rivers and lakes. Emerging new epidemics such as typhus

and cholera in London, in 1829 and 1831 respectively, followed. These diseases subsequently spread further throughout Europe and America (Arboleda, 2000).

Water–disease relationship
During the cholera epidemics of 1849 and 1854, in London, Dr John Snow discovered that the epidemic was caused by the pumping of contaminated water into wells and local water supplies (Tulodziecki, 2011). London promulgated the *Metropolitan Water Act* of 1852, requiring the filtration of all the supplied water to the city. Towards at the end of the 19th century, based on the discoveries of Louis Pasteur, Karl J. Eberth discovered the bacillus causing typhus in 1880, and Robert Koch discovered the bacillus Vibrio cholera in 1884. Therefore, the relationship between microorganisms present in water and the occurrence of diseases was demonstrated (Arboleda, 2000). Once the inter-relationship between contaminated water and disease became clear, the field of sanitary engineering developed rapidly. Large investments have been made in the physical infrastructure for water supply and wastewater collection and treatment ever since. Since then the engineering interventions have been based on two simple concepts (Gijzen, 1999; Harremoes, 2000): a) to break the transmission route of diseases by introducing 'filters' (drinking water treatment, chlorination, and later also wastewater treatment) in the urban water cycle; b) to transport human excreta out of the city (flush toilets and sewerage). These interventions combined with improved hygiene practices resulted in good public health and cities in industrialised countries have been essentially free of water-borne diseases since then (Bos *et al.*, 2004).

Technologies for wastewater treatment
Besides pathogens, effluents also contain other pollutants that can be detrimental to the receiving water quality and to the environment at large. Once the need to eliminate water pollutants before discharging into rivers had been recognized, a great interest in the development of technologies for wastewater treatment (WWT) started. Technologies for primary treatment arose between 1860 and 1914. Between 1914 and 1965 technologies such as activated sludge, artificial wetlands and rotating biological reactors were developed. From 1965 onwards, new regulations for the protection of the environment emerged in many countries (Lofrano and Brown, 2010). In this period, the emphasis was on more widespread application of known techniques for Biological Oxygen Demand (BOD) and Total Suspended Solids (TSS) removal; environmental protection and improvement of removal of nitrate, phosphate and ammonia nitrogen; and disinfection (mainly in the USA). Figure 2.1 shows the most significant developments in the evolution of wastewater treatment.

Water pollution control in the United States of America (USA)
In the USA, from about 1900 to the early 1970s, treatment objectives were concerned primarily with: 1) the removal of colloidal, suspended, and floatable material; 2) the treatment of biodegradable organics, and 3) the elimination of pathogenic organisms (Metcalf and Eddy, 2003). From the early 1970s to about 1980, wastewater treatment objectives were based primarily on aesthetic and environmental concerns. The earlier objectives, involving the

reduction of BOD, TSS and pathogenic organisms, continued, but at levels that were more stringent. Removal of nutrients, such as nitrogen and phosphorus, also began to be addressed, particularly in some of the inland streams and lakes, estuaries and bays (Metcalf and Eddy, 2003). The objectives for improvement of water quality that had been formulated in the 1970s were defined in more detail in the 1980s, emphasizing the removal of constituents that may cause long-term health effects and environmental impacts. Besides, in the last decades, it has become evident that diffuse sources of pollutants, including discharges from separate storm drainage systems and Combined Sewer Overflows (CSO) are major causes of water quality problems. In 1987, in responding to this situation Congress asked the United State Environmental Protection Agency (U.S. EPA), to regulate storm water discharges to protect water quality (U.S. EPA, 2008).

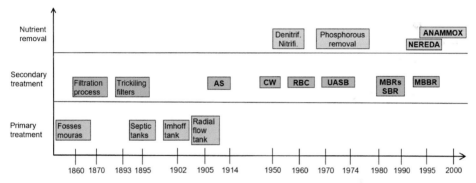

AS-Activated Sludge; CW-Constructed Wetlands; RBC-Rotating Biological Reactors; UASB-Upward Flow Anaerobic Sludge Blanket; MBRs-Membrane Biological Reactors, SBR-Sequencing Batch Reactors; MBBR-Moving Bed Biofilm Reactors; NEREDA-Process Wastewater Treatment with Aerobic Granular Biomass; ANAMMOX-Anaerobic Ammonium Oxidation.

Figure 2.1 Evolution of wastewater treatment in 19th and 20th centuries
Adapted from (Lofrano and Brown, 2010)

Water pollution control in Europe
In 1898, the Royal Commission on Sewage Disposal was created by the UK government. The eighth report of this commission (in 1912) had significant effects since it was concerned with the standards (and testing methods) to be applied to sewage and effluents being discharged to rivers. It recommended the so-called '20:30 standard', 'Royal Commission Standard' or 'General Standard'. This standard was copied by many other countries. This is a general standard allowing maximum values of 20 mg/L BOD and 30 mg/L SS in effluent discharges from WWTP (Cooper, 2001). These criteria were the basis for the design and operation of wastewater treatment plants. During the Second World War (1939-1945) there was limited progress on this issue. Subsequently, in the 50s, with the formation of the United Nations Organization and the increase in research into water and sanitation, the definition of water quality standards for different uses was introduced. In turn these advances contributed to the development of wastewater treatment technologies (Metcalf and Eddy, 2003; Lofrano and

Brown, 2010). Early European water legislation began, in a 'first wave', with standards for rivers and lakes used for drinking water abstraction in 1975 and culminated in 1980 in setting binding quality targets for drinking water. It also included legislation on water quality objectives for fishing waters, shellfish waters, swimming waters and groundwater (European Commission - DG Environment, 2008). At the end of 2000, the Water Framework Directive (WFD) was created, establishing a legal framework to protect and restore clean water across Europe and ensure its long-term, sustainable use. The WFD presented the following key aims: expand the scope of water protection to all waters, surface waters and groundwater; achieve 'good status' for all waters by a set deadline; water management based on river basins, a 'combined approach' of emission limit values and quality standards; getting the prices right; getting citizens more closely involved; and streamlining legislation (WISE, 2007).

Water pollution control in Latin America

Water pollution problems in Latin America and the Caribbean became evident during the 1970s. Most pollution was caused by agriculture and the discharge of untreated urban and industrial sewage. Agriculture contributes to the deterioration of surface water and groundwater by excessive run-off, soil erosion, fertilisers, herbicides, pesticides and organic waste. Excessive use of fertilisers in agriculture has increased eutrophication of lakes, dams and coastal lagoons (Savci, 2012). In Latin America, the wastewater treatment objectives proposed in 1970 related to the elimination of floating matter, biodegradable organic matter and pathogens. From the beginning of the 70s until the 80s, wastewater treatment objectives were more related to aesthetic and environmental criteria. However, starting in 1980, wastewater treatment systems began to address public health concerns arising from toxic substances present in wastewater discharges. In response to these treatment objectives and concerns, many countries in the region have introduced 'end-of-pipe' solutions (wastewater treatment plants WWTP) as pollution control measures. However, only about 14% of all effluents in Latin America and the Caribbean were receiving some kind of treatment at the end of the 20th century (WHO and UNICEF, 2000).

2.1.3 Water quality objectives versus wastewater treatment objectives

Control of water pollution has reached primary importance in developed and more recently also in many developing countries. The prevention of pollution at source, the precautionary principle and the prior licencing of wastewater discharges by competent authorities have become key elements of successful policies for preventing, controlling and reducing inputs of hazardous substances, nutrients and other water pollutants from point sources into aquatic ecosystems (Larsen *et al.*, 1997). To understand the different types of strategies in water quality management it is important to understand the following definitions and terminologies related to water quality and pollution control: 1) *Water quality criterion* (water quality guideline) is the numerical concentration or narrative statement recommended to support and maintain a designated water use. Water quality criteria are developed by scientists and provide basic scientific information about the effects of water pollutants on a specific water use; 2) *Water quality objective* is a numerical concentration or narrative statement, which has been established to support and to protect the designated uses of water at a specific site, river basin or part(s) thereof; 3) *Water quality standard* is an objective that is recognised in enforceable environmental control laws or regulations of a level of Government; 4) *Precautionary principle*, by virtue of which action to avoid the potential adverse impact of the release of hazardous substances shall not be postponed on the ground that scientific research has not fully proved a causal link between those substances, on the one hand, and the potential adverse impact, on the other.

Approaches to water pollution control initially focused on the fixed-emission approach and the water-quality criteria and objective approach. Lately, this emphasis has been changing in developed countries, while most developing countries continue applying the fixed-emission approach (Lofrano and Brown, 2010). Water quality objectives are derived from criteria by considering the local water quality, water uses, water movement, waste discharges, and socioeconomic factors. A major advantage of the water-quality objective approach to water resources management is that it focuses on solving problems caused by conflicts between the various demands placed on water resources, particularly in relation to their ability to assimilate pollution. Using the basin as the unit for analysis plays an essential role in setting water quality objectives. It provides the context in which the demands of all water users can be balanced against water quality requirements (Heinz *et al.*, 2007). Basin planning also provides the mechanism for assessing and controlling the overall loading of pollutants within whole river basins and, ultimately, into coastal zones and seas, irrespective of the uses to which those waters are put. The elaboration of water quality objectives and the selection of the final strategy for their achievement necessarily involves an analysis of the technical, financial and other implications associated with the desired improvements in water quality. The establishment of a time schedule for attaining water quality objectives is mainly influenced by the existing water quality, the urgency of control measures and the prevailing economic and social conditions. In some countries, a phased approach to establish water quality objectives is applied. This gradual introduction is probably also the best approach for developing countries (Enderlein *et al.*, 1997).

2.2 Twenty first century and the limitations of 'end-of-pipe' solutions

2.2.1 Population growth and urbanization

Currently, 54% of the world's population lives in urban areas with a monthly increasing rate of 5.5 million people. By 2050, over 70 per cent of the global population will be urban residents (United Nations, 2014). At the start of the 20th century, there were only 12 cities with 1 million or more inhabitants. By 1950, the number of cities with over 1 million people had grown to 83. In 2008, when reaching 50% urbanisation, there were more than 400 cities with over 1 million and 19 with over 10 million inhabitants. Another trend is observed in the geographical shift in urbanisation rates. In developed regions, typically over 70% of the population is urban, while this is significantly below 50% in most developing regions. However, more recently the urbanization rate has been increasing mainly in less developed countries. Projections show that between now and 2050, an additional 2.5 billion people will add to the growth of urban areas, with some 90 per cent of the increase happening in Asia and Africa (United Nations, 2015). In 2015, cities accounted for 60% of global drinking water consumption, 75% of global energy consumption, and 80% of global greenhouse gas (GHG) emissions (Crittenden, 2015). The trend towards more urbanized societies and the growing number of people living in large cities has huge implications for freshwater use and wastewater management (United Nations, 2014). This imposes special demands associated with water transport, water quality maintenance and the management of excess water from storm events, among other challenges. In general terms, urbanization processes imply increases in water requirements, wastewater and solid-waste generation. These effects must be managed to avoid and prevent water quality degradation. Urbanization also tends to degrade local watersheds and their surrounding areas through deforestation and increases in impervious areas (UNESCO - IHP, 2015; Seeliger and Turok, 2013).

2.2.2 Urbanization, impermeability and stormwater

The urbanization process changes the landscape and the flows of materials and energy in urban areas. Changes in the landscape and transport of runoff are particularly important with respect to surface runoff and its characteristics. The most visible consequence of urbanization is the increase of the coverage of land surface by construction, pavements, roads etc. creating impermeable ground, which strongly limits the possibility of water infiltration. During runoff processes, the rainwater is contaminated, which leads to contamination of the receiving waters. Stormwater can be transported by combined sewers, along with domestic and industrial wastewater, or by separate sewers that discharge to the nearest stream or lake. In combined sewers, high stormwater flows exceed the capacity of the pipeline and excess flow must be diverted by flow regulators as combined sewer overflows (CSO) to the nearest receiving water (De Toffol *et al.*, 2007). CSOs contain not only rainwater, but also untreated wastewater and sludge. Their direct discharge into receiving waters causes serious pollution problems (Marsalek *et al.*, 2008).

2.2.3 Wastewater treatment

In many cases, collected wastewater is discharged directly into the environment without any treatment (WWAP, 2017). A country's level of industrial and municipal wastewater treatment is generally a reflection of its income level. On average, high-income countries treat about 70% of the wastewater they generate, while that ratio drops to 38% in upper middle-income countries and to 28% in lower middle-income countries. In low-income countries, only 8% of industrial and municipal wastewater undergoes treatment of any kind (Sato *et al.*, 2013). These estimates suggest that approximately 80% of all wastewater produced globally is released to the environment without treatment (United Nations-Water, 2015). There also appears to be significant variability across different regions. In Europe, 71% of the municipal and industrial wastewater generated undergoes treatment, while only 20% is treated in the Latin American countries. In the Middle East and North Africa (MENA), an estimated 51% of municipal and industrial wastewater is treated. In African countries, the lack of financial resources for the development of wastewater facilities is a major constraint in managing wastewater, while 32 out of 48 Sub-Saharan African countries had no data available on wastewater generation or treatment (Sato *et al.*, 2013).

The treatment of wastewater and its use and/or disposal in the humid regions of high-income countries (e.g. North America, northern Europe and Japan) are motivated by stringent effluent discharge regulations and public awareness about environmental quality. The situation is different in high-income countries in drier regions (e.g. parts of North America, Australia, the Middle East and southern Europe), where treated wastewater is often used for irrigation, given the increasing competition for water between agriculture and other sectors. The persistent expansion of sewerage and the consequent increases in wastewater volume generates pressure on existing treatment facilities, and in some cases can lead to suboptimal performance. Even when wastewater is collected and treated, the final quality of the wastewater discharged may be affected by poor operation and maintenance, as well as overflow during storm events, when wastewater is allowed to bypass the treatment plant. Thus, much of the wastewater is not treated (or inadequately treated) and discharged in water bodies, and subsequently affects the water quality (and its availability) for users downstream (WWAP, 2017).

2.2.4 Diffuse pollution in urban and rural areas

Diffuse sources of pollution are indirectly discharged to receiving water bodies, via overland and subsurface flow and atmospheric deposition to surface waters and leaching through the soil structure to groundwater during periods of rainfall and irrigation. The most severe water quality impacts from diffuse source pollution occur during storm periods (particularly after a dry spell) when rainfall induces hillslope hydrological processes and runoff of pollutants from the land surface (Bravo-Inclán *et al.*, 2013). More than 600 chemical pollutants have been identified in stormwater. These chemicals could affect human health and aquatic life. The list of contaminants associated with diffuse pollution includes solids, chloride, nutrients (N and P), pesticides, polycyclic aromatic hydrocarbons, bacteria, heavy metals, etc. (Marsalek *et al.*, 2008). Typical examples of diffuse pollution include the use of fertiliser in agriculture and forestry, pesticides from a wide range of agricultural land uses, contaminants from roads and paved areas, and atmospheric deposition of contaminants arising from industry (Environment Agency, 2007).

In Latin America, the challenges of water have been mainly focused on achieving water coverage and basic sanitation. The control of contamination by diffuse sources is practically ignored (Bravo-Inclán *et al.* 2013). After decades of regulation and investment to reduce point source water pollution, OECD countries are still facing water quality challenges (e.g. eutrophication) from diffuse agricultural and urban sources of pollution, i.e. pollution from surface runoff, soil filtration and atmospheric deposition. The relative lack of progress reflects the complexities of controlling multiple pollutants from multiple sources, their high spatial and temporal variability, the associated transaction costs, and limited political acceptability of regulatory measures. Reducing the costs of diffuse pollution requires much greater attention from policy makers. For OECD countries, the cost of current pollution from diffuse sources exceeds billions of dollars each year. These costs are associated with: the degradation of ecosystem services: water treatment and health-related costs; impacts on economic activities such as agriculture, fisheries, industrial manufacturing and tourism. The scale of these costs means that seeking increasingly marginal reductions in point source pollution is no longer the most cost-effective approach to improving water quality in many OECD countries (OECD, 2017).

2.2.5 Micropollutants

Micropollutants (MPs) include organic or inorganic substances with persistent, bio-accumulative and toxic properties, which may have adverse effects on human health or/and biota. MPs can be considered as persistent organic pollutants (POPs) if their physical and chemical properties remain intact for long periods once they are released into the environment. POPs accumulate in the adipose tissues of living organisms including humans. Higher concentrations of POPs have been found in food chains exposing humans and wildlife to toxic effects and diseases (Fürhacker *et al.*, 2016). Thousands of chemicals play an important role in our daily activities. As a result of widespread use, these substances also enter the environment. A significant pathway for the input and spread of chemicals is water - for example, when substances are washed out by rainwater or transported by wastewater (Wittmer and Burkhardt, 2009). Grey water, which originates from the kitchen, bathroom or laundry, can contain over 900 synthetic organic compounds or xenobiotic (Erikson, 2002). Residues of pharmaceuticals after use by humans enter raw sewage via urine and faeces and by improper disposal. These pharmaceuticals are discharged from private households and hospitals, and eventually reach municipal WWTPs. Many of these pharmaceutical residues and hormones (anti-conception drugs) are recalcitrant compounds that are not efficiently broken down and therefore end up in receiving waters even after WWT. Some of these compounds may undergo microbial transformation into products with even higher human and eco-toxicological behaviour. The presence of pharmaceuticals and oestrogenic compounds in natural and drinking water has indeed been reported in recent years (Navarro and Zagmut, 2009).

2.2.6 Climate change and impact on water resources

The earth's energy balance determines the functioning of its climate system, depending on a number of factors. Some of these factors are natural, such as variations in solar energy, and some are anthropogenic in origin, such as the changes in the quantity of greenhouse gases (GHGs) in the atmosphere (Posada, 2008). Carbon dioxide is the main GHG released by anthropogenic activities; others include methane and nitrous oxide. Increases in GHGs in the atmosphere avoid the release of thermal infrared radiation into space, leading to increases in the earth's surface and atmosphere temperature (Loftus, 2011). Climate change will lead to sea level rise and the intensification of the hydrological cycle, producing more frequent and intense rainfall as well as extended dry periods. As a result, a city's water supply, wastewater and stormwater systems will be particularly affected. Climate change impacts on the urban water system typically have knock-on effects on other urban systems because of the role that water plays in many urban processes and quality of life (Novotny, 2008). One of the most important findings of Bates *et al.* (2008) has been the linkage between the global warming observed in recent decades and the large-scale changes in the hydrological cycle, such as changes in vapour content in the atmosphere, precipitation patterns, rainfall intensity and frequency of extraordinary storms, snowpack depth, glacier cover, soil moisture, and runoff processes. In many lakes and reservoirs of the world, climate-change effects are mainly due to variations in water temperature affecting oxygen regimes, oxidation/reduction (redox) reactions, stratification, mixing rates, and the development of biota (Montes-Rojas *et al.*, 2015). For example, increasing the temperature decreases the self-purification capacity of rivers by reducing the amount of dissolved oxygen, which in turn limits biodegradation. An increase in heavy precipitation leads to increased nutrients, pathogens, and toxins in water bodies (Kundzewicz *et al.*, 2007).

2.2.7 Other sources of stress on water systems

Water resources are substantially affected by human activities such as dam building, deforestation, mining activities, land use changes and pollutant loads. Human activities can exacerbate the negative impacts of climate change by increasing the vulnerability of systems to a changing climate (Bates *et al.*, 2008). Other impacts are associated with house construction in sensitive areas, such as on high slopes in the upper parts of water catchment areas, and close to sensitive groundwater aquifers. In recent decades, the growing damage to freshwater resources coincides with the increased demand for water. The erosion associated with deforestation has also altered the (local) water cycle and has caused the loss of soil, increasing the sediment load transported towards the coasts.

2.2.8 Limitations of 'end-of-pipe' solutions

The protection of water resources from quality deterioration by point and non-point source pollution discharges is probably the biggest challenge in sustainable water resources management over the coming decades. In the 60s and 70s we started to see the first signs of the 'pandemic' of water pollution. In practice, most countries adopted pollution-control approaches which were based exclusively on 'end-of-pipe' solutions by constructing WWTPs. The results have shown that this strategy has not fully complied with the planned objectives. Continuing the urban water practice in this 'business-as-usual' manner is very costly, unsustainable, and is leading to significant problems related to public health, water quality and the environment at large and, thus, the economy. In order for investments in water and sanitation to produce the expected outcomes in quality of life improvement in communities, a holistic vision of the problem is necessary. To achieve this goal, water management must consider the basin as the unit of analysis and consider water quality objectives in the planning of investments in pollution prevention and control.

In the last few decades, an important number of concepts related to sustainable water management have emerged. Additionally, different methods and innovative approaches have been published that use these concepts and propose alternatives to the business-as-usual practise of end-of-pipe treatment. The following section gives an overview of some of these innovative approaches, while in Section 2.4, the Three-Step Strategic Approach (3-SSA) is presented.

2.3 Water in crisis: some concepts to sustainable water management

2.3.1 Water sustainability

A short definition of 'sustainable development' was presented by the Brundtland Report: 'development that meets the needs of the present without compromising the ability of future generations to meet their own needs' (WCED, 1987). This report adopts the definition of 'water sustainability' by which water resources and water services are able to satisfy the changing demands placed on them, now and into the future, without system degradation (ASCE, 1999). It also incorporates the four Dublin Principles (1992): 1) freshwater is a finite and vulnerable resource, essential to sustain life, development and the environment; 2) water development and management should be based on a participatory approach, involving users, planners and policy-makers at all levels; 3) women play a central part in the provision, management and safeguarding of water; 4) water is a public good and has a social and economic value in all its competing uses (GWP, 1992).

2.3.2 Resilience

The concept of resilience has evolved over the past 40 years, arising from a narrow perspective with specific applications (engineering resilience) to a broader perspective that encompasses a more comprehensive application context (socio-ecological resilience). The resilience concept has modified existing views that considered the systems stability as an imperative, by introducing a new perspective that considers the capacity of systems to adapt and change. This increases the probability of sustainable development in changing environments where the future is unpredictable (Blanco et al., 2017). Based on concepts of Holling (1973) and Walker et al. (2004), resilience was defined as 'the potential of a system to tolerate disturbances without collapsing towards a qualitatively different state, maintaining its structure and function, which implies its capacity to reorganize itself, following the changes driven by disturbances'. Socio-ecological resilience is characterized by the interactions between disturbances, reorganization, recovery, sustainability and development in a system, and it depends on the capacity to adapt, transform, learn and innovate in a context of unstable equilibrium (Folke, 2006; Blanco et al., 2017).

2.3.3 Integrated Water Resources Management

Integrated water resources management (IWRM) is a systematic process for the sustainable development, allocation and monitoring of water resource use in the context of social, economic and environmental objectives (GWP and SAMTAC, 2000). IWRM defines the basin as a planning unit. The hydrographic basin is defined as a natural system composed of several components: 1) the biophysical formed by water and air; 2) the biological formed by the flora, the fauna that is found in a terrestrial, aquatic ecosystem. The boundaries of the systems are established by the watershed from the precipitated water input to the total water output (Castro, 2008; Ordóñez, 2011). The basin scale allows the analysis of the interactions between the cities and the basins in which the cities are embedded. Activities at the catchment scale include flood protection and facilitate the implementation of strategies to allow access to adequate quality water. Upstream changes in land use patterns or water allocation may change the local hydrology and available water resources and can result in the necessity of basin protection plans or water allocation strategies (Gleick, 2009; Anderson and Iyaduri, 2003). On the other hand, the city's impact on the watershed has to be considered. This may refer to the efficient use of the water resources within cities as well as the impact of cities on downstream uses through the discharge of wastewater and storm water (Vairavamoorthy et al., 2015). The application of IWRM requires effective governance. This process is characterized by the participation of the stakeholders, building a shared vision of sustainable development, able to satisfy human requirements without damaging the natural resources. The hydrological complexity and the limited participation of users and government entities may hinder the effectiveness of IWRM (Tejada-Guibert, 2015).

2.3.4 Urban water cycle

The hydrological cycle determines the storage and circulation of water among the biosphere, atmosphere, lithosphere, and the hydrosphere. Combined effects of urbanisation, industrialisation, and population growth affect natural landscapes and the hydrological characteristics of watersheds (Marsalek et al., 2008). The hydrological cycle becomes more complex in urban areas, because of many anthropogenic influences and interventions, giving way to the urban basin concept. Therefore, the 'urban' hydrological cycle is usually referred to as the Urban Water Cycle (UWC). Urban basins are characterized by settled populations or urban areas through their ecosystem services such as supply (water, energy, food and raw materials, transport, communication), regulation (climate, erosion, diseases), cultural construction (recreation, landscape aesthetics, education, cultural heritage) and support (soil formation, biodiversity of productive activities), among others. The UWC is the spatiotemporal interaction between water and hydrological processes, as well as supply, treatment, distribution, consumption, collection, provision, and reuse carried out in urban areas (Marsalek et al., 2008; Wagner and Breil, 2013). The UWC is modified by external and internal factors. These factors intervene both directly and indirectly within each input, thereby increasing the entire cycle's complexity (Table 2.1).

Table 2.1 Internal and external factors of the Urban Water Cycle

UWC Part	UWC Components	Internal Factors	External Factors
Water supply sub-system	Raw-water intake	Population, availability, techniques	Climate, environment, economy, geography
	Water treatment	Population, techniques, quantity, quality, energy	Climate, economy, regulations, geography
	Storage	Population, techniques, energy	Climate, environment, economy, geography
	Water supply distribution	Population, techniques, quantity, quality, energy	Economy, geography, society, culture, environment, regulations
Water demand	Water consumption	Population, weather, population density, land use, equipment, economy	Education, territory growth, culture, regulation
Wasterwater and stormwater subsystem	Collection	Population, weather, population density, land use, equipment geography, hydraulics, regulations, public health, environment, economy	Society, culture, education
	Treatment	Land use, equipment, geography, regulations, public health, quality, quantity, environment, economy, energy	Society, culture, education
	Receiving water	Equipment, geography, regulations, public health, quality, quantity ecology, environment, economy	Territory growth, type of water-receiving body

Source: (Peña-Guzman *et al.*, 2017)

2.3.5 Integrated Urban Water Management

The concept of Integrated Urban Water Management (IUWM) was founded on the premise that the design and management of the urban water system is based on the analysis and optimization of the whole system, rather than the analysis of individual tasks related to urban water services and resources (Van der Steen and Howe, 2009). An integrated approach for the design and urban water management would provide opportunities for more efficient and sustainable use of water resources. These opportunities include: 1) taking advantage of stormwater (urban rainwater harvesting) and treated wastewater effluent as alternative water supply sources; 2) the control of stormwater quality and quantity to achieve more efficient wastewater treatment through combined sewer systems; 3) the use of aquatic ecosystems for water purification and natural flood protection purposes. IUWM includes different urban sectors such as land use, housing, energy and transport. Likewise, this approach considers other non-urban uses of water resources, recognizes the population, local authorities and other stakeholders that govern the cities, and seeks economic equilibrium, social equality and environmental sustainability (GWP, 2012). Progressively, IUWM has diversified and also integrated recreational and aesthetic water uses as well as pollution control to preserve the ecological flows and the natural geomorphological characteristics (Figure 2.2).

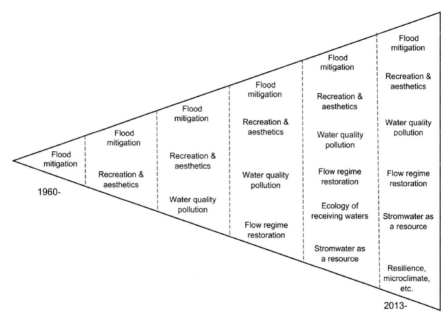

Figure 2.2 Increasing integration and sophistication of urban drainage management over time (Fletcher et al., 2015)

2.3.6 Household-Centered Environmental Sanitation approach

The Household-centred Environmental Sanitation (HCES) approach was conceived by the Environmental Sanitation Working Group of the Water Supply and Sanitation Collaborative Council (WSSCC) in 1999, for people in developing countries. The HCES approach is a radical departure from past central planning approaches as it places the household and its neighbourhood at the core of the planning process. The approach responds directly to needs and demands of the users, and it attempts to avoid problems resulting from purely 'bottom-up' or 'top-down' approaches. It offers the promise of overcoming the shortcomings of unsustainable planning and resource management practices of conventional approaches (Morel *et al.*, 2003). According to the HCES approach, the way to reach sustainable solutions includes the following:

Water demand management: to minimize wasteful use of water, reduce the need for new source and production of wastewater.
Reuse and recycling of water: minimize the wastewater collection, treatment and disposal.

Solid waste recycling: reduce burden of collection and disposing of solid waste.

Nutrient recovery: either at household level (eco-sanitation) or on a wider scale (urban agriculture).

Improved rainwater management: including detention and treatment, and reuse of stormwater.

Strong emphasis on intermediate technologies: encourage household and community-level construction, operation and management of facilities, and permit reuse and/or disposal at local level.

Institutional arrangements and mechanisms: encourage the participation of the private sector, facilitate cooperation across zone or sub-zone boundaries and ensure the provision of technical assistance.

Economic analysis procedures: economic benefits of good planning as well as the consequences of sub-optimal desirable alternatives.

Effective and sustainable financial incentives: that determine whether problems should be solved within the zone itself, or whether a joint solution should be selected to serve more than one zone.

Cost recovery practices: that ensure financial viability, social equity; promote the 'circular system' and the productive use of the waste.

2.3.7 Ecohydrology

Ecohydrology is a trans-disciplinary approach, using the understanding of relationships between hydrological and biological processes at the catchment level to improve water quality, biodiversity and sustainable development (Zalewski, 2006; Wagner *et al.*, 2007). The implementation of this approach is based on the restoration and maintenance of water circulation patterns, nutrient cycles and energy flows at a catchment scale towards optimization of the ecosystem services for society (Zalewski and Wagner, 2008). The main areas for eco-hydrology applications include the following: 1) Increasing the water catchment, retention and flow duration through the maintenance of existing forest cover, reforestation, and wetland protection; 2) decreasing the loading of non-point pollution by soil conservation and maintaining riparian vegetation along stream courses; 3) maintaining in-stream habitat by the maintenance/restoration of natural river channels and floodplains, ensuring a natural seasonal flow regime; 4) employing biogeochemical processes in natural and constructed wetlands to treat organic matter and nutrient-laden sewage (Saha and Setegn, 2015). A watershed planning and management strategy within a hydrologically defined area provides a coordinated framework for water supply protection, pollution prevention, and ecosystem preservation. Although watershed strategies vary, they should be based on an integrated study of ecosystems and hydrological characteristics, processes and their combined potential to influence water dynamics and quality. Ecohydrology requires an understanding of temporal and spatial patterns of catchment-scale water dynamics, which are determined by four fundamental components: climate, geomorphology, plant cover/biota dynamics, and anthropogenic modifications (UNEP, 2003).

2.3.8 Water governance

Water governance is related to the political, social, economic and administrative systems responsible for water resources development, management and service delivery at different levels of society (Rogers and Hall, 2003). Hence, the essence of water governance is more related to how decisions are made (i.e. how, by whom, and conditions for decision-making) than the decisions themselves (Moench *et al.*, 2003). Water governance includes the application of policies and regulations for the water and other natural resources management, involving the formal and informal institutions by which authority is exercised. Effective water governance is required for success in IWRM, but in many countries, it is not working well. Water managers can promote good water governance, especially by implementing effective management practices and promoting productive relationships among stakeholders (Grigg, 2016).

2.4. Three-Step Strategic Approach

2.4.1. The concept of Three-Step Strategic Approach

The Three-Step Strategic Approach (3-SSA) is based on cleaner production principles and lessons learned from its long-time application in industry. The 3-SSA provides new alternatives to the limited and unsustainable achievements provided by end-of-pipe solutions (Naphi and Gijzen, 2005; Gijzen, 2006). The three steps include: 1) prevention or minimisation of waste production; 2) treatment, recovery and reuse of waste components, and 3) disposal of waste with stimulation of natural self-purification in the receiving water body (Figure 2.3). For maximum benefit, the steps should, preferably, be implemented in chronological order, and possible interventions under each step should be fully exhausted before moving on to the next step (Nhapi and Gijzen, 2005).

2.4.2 Cleaner production concepts in Urban Water Management

Cleaner production concepts have been successfully applied in the industrial sector. Gijzen proposed that the cleaner production concept, successfully developed for industries over the past decades, could also help to transform urban water management (Gijzen, 2001a, 2001b, 2006). Table 2.2 demonstrates that there is a sharp contradiction between the identified cleaner production principles and current water management practices. These cleaner production principles were the basis for formulating the 3-SSA (Gijzen, 2006).

2.4.3 Step 1. Prevention or minimisation of waste production

At the household level
In the urban water context waste minimization can be achieved via three main actions (Gijzen, 2006; Cardona, 2007; Galvis *et al.*, 2014): 1) reduction at source, which includes a change in consumption habits and application of low consumption devices; 2) in-situ recycling techniques, and 3) rainwater harvesting. The first category of actions proposes a shift to *low*

consumption devices, such as water-saving toilets, showers and aired faucets able to generate decreases in water consumption, allowing for the possibility of supplying more users, without the need for additional water sources and treatment capacity. The second and third categories of actions recognize new alternative water sources, such as rainwater harvesting and grey water. The *use of treated grey water* is feasible for toilet flushing, plant watering, and the washing of floors and outdoor areas (Mejia *et al.*, 2004; Gijzen, 2006; Sierra, 2006; Liu *et al.*, 2010), golf courses, agriculture and groundwater recharge (Ottoson and Stenström, 2003).

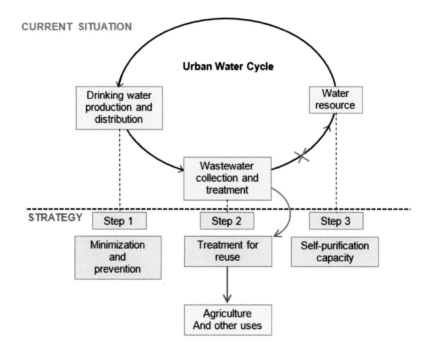

Figure 2.3 Schematic representation of the 3-Step Strategic Approach to wastewater management. Adapted from Nhapi and Gijzen (2005)

At the urban drainage system level

Most of the existing drainage systems have one or more of the following problems: negative impacts on receiving water by storm water runoff discharges, runoff pollution, dilution of influent to the WWTP, discharges from combined sewer overflows (CSOs) or illegal connections to sanitary sewer systems (Marsalek *et al.*, 2008). The high rates of urbanisation are contributing to the increase in impervious areas and runoff volume. The impervious areas have caused localised flooding, and water pollution. These effects have increased with climate changes. The conventional systems for stormwater management have not been an efficient solution. This calls for a change in stormwater management avoiding water cycle interruptions and allowing its use and storage. Sustainable Urban Drainage Systems (SUDS) provide this option. SUDS are aimed to reproduce the natural water cycle as closely as possible to how it existed prior to urbanization (Novotny and Brown, 2007). SUDS maximize the opportunities and benefits that can be achieved from storm water management (Mitchell, 2006; Fletcher *et al.*, 2015). Examples of SUDS options include the following: green roofs, soakaways, rainwater harvesting, filter strips, trenches, swales, bio-retention, pervious pavements, infiltration basins, detention basins, ponds and wetlands (Bregulla *et al.*, 2010, CIRIA, 2015).

Table 2.2 Cleaner production principles and current water management practices

Principle	Practice
Use lowest amount of input material, energy or other resources per unit of product	We supply between 130 and 350 L of drinking water per capita per day, while less than 2 L are actually used for drinking.
Do not use input materials of a higher quality than strictly necessary?	We use water purified to drinking water standards to flush toilets, clean floors, wash cars and irrigate the garden.
Do not mix different waste flows	Various wastewater flows are already combined (urine and faecal matter, grey and black water) in the household. After disposal into the sewer this combined waste is mixed further with industrial effluents, and often also with urban runoff. Obviously this practise makes re-use of specific components in the mixed waste flow less attractive and less feasible.
Evaluate other functions and uses by products before considering treatment and final disposal	Domestic sewage is discharged into open water resources either with or without prior treatment. Only a few examples of wastewater re-use or (by-) product recovery from wastewater exist.

Source: (Nhapi and Gijzen, 2005; Gijzen, 2006)

2.4.4 Step 2. Treatment, recovery and reuse of waste components

The second step of the 3-SSA focuses on treatment technologies for wastewater reuse. The reuse of effluents becomes beneficial if there is a demand, and the requirements in water quality standards can be achieved through cost-effective treatments (Vairavamoorthy *et al.*, 2015). Potential benefits include: 1) savings on water use, as the use of treated effluents will reduce the use of freshwater resources in activities such as crop irrigation, industrial processes, cleansing or washing activities. An interesting example is the so-called 'new water' concept in Singapore, where sewage is treated to generate safe drinking-quality water (Tortajada, 2006; Public Utilities Board, 2016). Besides wastewater reuse being an additional source of water, it also generates other environmental and economic benefits by allowing more water availability for sensitive ecosystems and recreational activities; 2) savings in fertilizer use by the reuse of effluents and bio-solids. Effluent reuse improves soil productivity contributing organic matter and macronutrients (N, P, K), thus reducing the use of chemical fertilizers (Hespanhol, 2003; Corcoran *et al.*, 2010; Winpenny *et al.*, 2013); 3) reduction in sewer tariffs and taxes for wastewater discharges directly to water bodies. The reduction of effluent discharges contributes directly to an improvement of the water quality of the receiving water bodies (Bixio and Wintgens, 2006); 4) Converting the Chemical Oxygen Demand (COD) into energy. Wastewater can be treated in aerobic or anaerobic systems, but anaerobic systems appear to be more favourable because of energy recovery in the form of CH_4, which contributes to cost-effectiveness (Gijzen, 2001a). When organic matter is anaerobically treated, about 375 L of methane can be produced per kilogram of BOD digested. Assuming almost complete conversion of organic matter present in sewage into biogas, a daily production of 25 to 45 L of methane per capita can be expected (Nhapi and Gijzen, 2005); 5) Savings on irrigation infrastructure and its operation and maintenance (O&M), when groundwater is used for irrigation. With wastewater reuse in agriculture, groundwater is preserved (Moscoso *et al.*, 2002). In addition to agricultural reuse, infrastructure costs and pumping groundwater may be avoided (Cruz, 2015). With agricultural reuse, freshwater from surface and underground sources remains available for water supply or other ecosystem services (Winpenny *et al.*, 2013).

The challenge is to develop adequate treatment systems that produce biologically and public health-safe effluents, preserving the valuable components such as nutrients, which may replace fertilizers (Regmi *et al.*, 2016). It is important to understand that measures under Step 1 would lead to a smaller volume of more concentrated wastewater, which allows for other WWT technologies under Step 2 to become viable, such as anaerobic wastewater treatment (AWWT) with biogas recovery. In 1989, the World Health Organisation (WHO) developed guidelines for the safe use of wastewater in agriculture. The 2006-updated version was the result of gathering new epidemiological evidence and the use of quantitative microbial risk assessment (WHO *et al.*, 2006). Currently many countries do not have their own guidelines on the use of treated wastewater. Additionally, there is a limited knowledge about treatment technologies to ensure the quality of effluent treatment systems according to the different types of reuse. This has stimulated the use of raw wastewater.

2.4.5 Step 3. Disposal of remaining waste with stimulation of natural self-purification

Once the steps 1 and 2 have been fully exhausted it may be necessary to resort to Step 3 if some unmanaged contamination still remains in the effluent, and no reuse option can be found. Step 3 aims to reduce pollutant concentrations and exposure risks by promoting natural self-purification in receiving water bodies. Usually the local environment suffers initially after receiving effluent discharges, and therefore one strategy is to boost the self-purification capacity of the receiving water body so that it can cope with the pollution load (Gonzalez *et al.*, 2012). In the 3-SSA, the essence of Step 3 is to help stimulate this natural self-purification capacity. This can be achieved via simple ecohydrology interventions. Another complementary strategy is to use the self-purification assessment of water bodies to guide decisions regarding effluent treatment levels and discharge points.

Aquatic ecosystems have an inherent capacity to maintain water quality that is referred to as the overall assimilative capacity of a particular stream, river, or wetland (e.g., McClain, 2008). Ostroumov (2005 and 2006) reported the array of physical, chemical, and biological processes that contribute to maintaining water quality. Physical processes include filtration, deposition, and dilution. Chemical processes include sorption/release of substance from sediments and organic matter and transformation through biogeochemical reactions. Biological processes include sequestration, microbial transformation, uptake by plants and animals, and nutrient spiralling. These processes are interconnected and depend upon the existence of different habitat types and zones such as streams, floodplains, and riparian vegetated zones (Saha and Setegn, 2015). Chemical and biological changes that occur in a river downstream of a sewage discharge point are illustrated in Figure 2.4. Microbial processes in rivers are responsible for the degradation of the organic components. Since oxygen is consumed by aerobic microbial biodegradation, its level drops over a certain distance from the waste discharge point (the 'oxygen sag curve') until it starts recovering again due to biodegradation and the re-aeration process. The decrease in oxygen is followed by an increase in nutrients, which results from the mineralization of the organic matter. Uptake of these nutrients by algae and water plants is responsible for the subsequent decrease further downstream (UNEP *et al.*, 2004). These algae and water plants may subsequently contribute to supplying additional oxygen to the water body during daytime photosynthesis, thereby contributing further to stimulating the self-purification capacity.

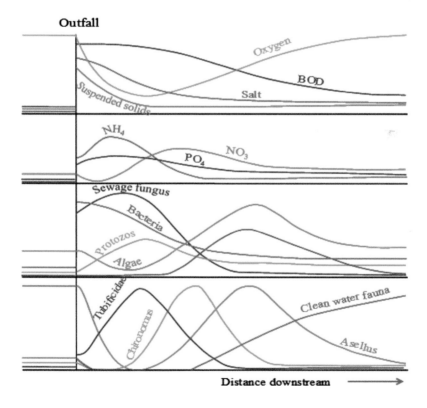

Figure 2.4 Changes in a river downstream of a sewage outfall
(Hynes, 1960)

The self-purification capability is understood as a process in which different mechanisms act, helping to assimilate or transform organic and inorganic matter. The interaction of these mechanisms is complex and has been studied through mathematical models. The first models were developed by Streeter and Phelps in 1925. In rivers, the self-purification capacity depends mainly on: 1) hydraulic characteristics of the receiving water (flow, depth, wetlands, floodplains); 2) quality of the receiving water, diluting the pollution discharges and facilitating subsequent (bio)degradation of organic matter; 3) water turbulence, which provides oxygen to the water favouring microbial biodegradation activity, and 4) the nature and size of the discharges (Vagnetti, 2003; Von Sperling, 2005). The self-purification capacity will largely depend on dissolved oxygen (DO) levels. Once a water body turns anoxic, the self-purification capacity is decimated. Other factors include the presence and activity levels of algae and aquatic plants, which may enhance microbial decomposition processes during the daytime due to higher DO levels.

These phenomena can also be pro-actively stimulated via targeted ecohydrology interventions. The natural purification capacity of receiving water bodies can be encouraged by allowing rivers to flow outside their often artificial embankments (Boraschi, 2009; García-Quiroga and Abad-Soria, 2014). The generated floodplains and wetland surface area will contribute in terms of self-purification of the water body, mainly due to prolonged retention time and improved aeration by algae and wetland plants, particularly in the shallower areas of the water body. Other options include the construction of small dams to cause rapids and turbulence in streams for improved aeration of the river water. This will boost the aerobic heterotrophic activity of bacteria in the water. Also the introduction or stimulation of controlled algal development to stimulate oxygenation could be considered (Zalewski, 2000). An example of stimulating natural self-purification capacity is the heavily polluted 'Bocana de La Virgen' Bay, in Cartagena, Colombia (Moor et al., 2002; Gijzen, 2006).

Six inlet and four outlet doors were constructed to allow water inflows and effluent outflows to be controlled by tidal pressure. This action improved the water quality as dilution occurred and self-purification was enhanced. The inherent capacity of a particular water body is assessed by ecohydrologists; it is then used along with a factor of safety by water managers to determine the total daily maximum loads (TMDLs) of pollutants in discharges by different point sources along with prevailing nonpoint sources. By recognizing and stimulating the self-purifying functions that a natural stream or river provides, water quality can be maintained at source which vastly decreases the expense of treatment at the user's end. In most developing countries, maintaining good water quality in streams, rivers, and wetlands is the only way to ensure water quality, given the unfeasibility and unsustainability of large treatment plants (Saha and Setegn, 2015).

2.5 Technology selection for water quality management

2.5.1 Criteria and methodological tools

The decision making to select technological processes is usually accompanied by methodological tools that facilitate the process. This process includes multiple aspects or criteria, from the technical, environmental, social and economic points of view, in order to increase the sustainability of the technology implemented. Most of these processes use economic models and/ or multi-criteria type of models. The models that consider water quality objectives (see Section 2.1.3) are based on mathematical modelling.

Cost Benefit Analysis (CBA) is a technical evaluation that allows the convenience and opportunity of a project or a solution alternative to be defined, comparing the Net Present Values (NPV) of the costs and benefits (Miranda, 2000). The objective of implementing the CBA is to weigh the positive and negative effects of an investment decision, which can be manifested internally or externally to the solution formulated. In this type of analysis, the benefits of the proposed action are calculated and compared with the total costs that society would assume if the said action were to be carried out (Brent, 2006). A variant of the CBA is when the 'incremental' situation is considered. An 'incremental analysis' of CBA is a decision-making technique used in business to determine the true cost difference between alternatives. It is also called the relevant cost approach, marginal analysis or differential analysis. 'Incremental' means that common benefits and common costs are not considered. The discount rate for net present value (NPV) is an efficiency criterion used in CBA to cases where costs and benefits occur over time. The discount rate corresponds to the return that could be earned per unit of time on an investment with similar risk. The social discount rate (SDR) is the rate used in computing the value of funds spent on social projects (Harrison, 2010).

Multi-criteria decision analysis (MCDA) is concerned with structuring and solving decision and planning problems involving multiple criteria. The purpose is to support decision-makers facing such problems. Typically no unique optimal solution for such problems exists and it is necessary to use the decision-maker's preferences to differentiate between solutions (Hajkowicz and Collins, 2007). Water-resource management decisions are typically guided by multiple objectives measured in different units. Multiple criteria decision analysis (MCDA) represents a body of techniques potentially capable of improving the transparency, auditability and analytic rigour of these decisions (Dunning *et al.,* 2000). The MCA framework ranks or scores the performance of alternative decision options against multiple criteria which are typically measured in different units. MCA emerged as a decision analysis technique in the 1960s and 1970s, partly resulting from the rapid growth of operation research. Water management is typically a multi-objective problem which makes MCDA a well-suited decision support tool (Hajkowicz and Collins, 2007). Whilst selection of the MCDA technique is important, more emphasis is needed for the initial structuring of the decision problem, which involves choosing criteria and decision options (Hajkowicz and Higgins, 2008; Mutikanga *et al.,* 2011).

River water quality modelling. With the development of model theory and the rapid increase in computer capacities, more and more water quality models have been developed for different topography, water bodies, and pollutants at different space and time scales (Liou *et al.*, 2003; Wang *et al.*, 2004). The historical development of these models has been closely linked to the type of parameter and the knowledge of its behaviour in the system to be modelled: dissolved oxygen and organic matter, total phosphorus, nutrients and phytoplankton, pathogen microorganisms, toxic organic compounds and heavy metals, sediment solids and sludge, acid rain and bio-accumulation (Monerris and Marzal, 2001). Most of the simulation river models have a similar conceptual model of the hydrodynamic and quality components.

The development of surface water quality models started in 1925, when Streeter and Phelps implemented an analytic expression to determine the oxygen content throughout a river exposed to a continuous discharge of biodegradable organic matter. In 1954 Campe developed a model increasing the amount of variables considered in the Streeter and Phelps model (Von Sperling, 2005). In 1969, Vollenweider developed an analytic solution for the calculation of the total phosphorus concentration in the water of a lake. In 1971, the SWMM model (Block Receiving) was developed for the U.S. EPA by the University of Florida, Metcalf and Eddy and Water Resources Engineering (Galvis *et al.*, 2006). In 1973, the QUAL2 model for rivers appeared and afterwards, new versions such as the QUAL 2E (1987) and QUAL 2K (2003) appeared. The WASP model for rivers, estuaries, lakes and coastal areas appeared in 1983. A broad development of multi-dimensional models took place in the 1980-1990 decade, accompanied by the development of numeric techniques for the solution of general equations, development of stronger and more robust computer equipment, progress in the investigation of the behaviour of substances and the beginning of commercialisation of certain computer packages for PCs and workstations (Monerris and Marzal, 2001). Ecosystem models have been used during the last few decades. These can show suspended solids, diverse algae groups, zooplankton, invertebrates, plants and fish (Von Sperling, 2007).

There are one-dimensional models for rivers (e.g., SWAT, MIKE 11 and QUAL-2K), two-dimensional models for lakes and reservoirs (e.g., CE-QUAL-W2, MIKE 21), and three-dimensional models for estuaries (e.g., WASP and ELCOM-CAEDYM). With the increasing importance of water quality, more and more elements are being included in models to assist in studying and managing water quality. However, there are a number of challenges, among which are the information requirements and the need to have reliable validation processes (Wang *et al.*, 2013).

2.5.2 Technology selection models for water quality management

Initially, the development of technology selection methodologies for water quality management was mainly oriented to its application in developing countries (Hamouda, 2011). For developing countries, the common selection criteria identified by different authors can be classified into the following factors: treatment objectives, technological aspects, costs, operation and maintenance, wastewater characteristics, demographical and socio-cultural factors, site characteristics, climate factors, environmental impact, capacity and willingness to pay, and construction aspects (Galvis *et al.*, 2006; Singhirunnusorn, 2009; Hamouda, 2011). Apart from the mathematical optimization approaches, there are other methods ranging from non-mathematical and simple approaches, such as simple flow charts, to intelligent tools and computerized decision support systems. Below follows a brief description of some of the main types of the available methodologies.

Descriptive methods. These methods include basic universal guidelines for the selection of WWTP technologies, without referring to a specific scenario. They suggest starting by defining the requirements with which the final WWTP effluent must comply before proceeding to select the units or combination of processes needed to obtain an effluent with the required quality (Metcalf and Eddy, 1991 and 1995; Crites and Tchobanouglous, 1998).

Methodologies based on algorithms and check lists. UNEP (1998) proposed a decision tree considering different alternatives, such as individual systems, site disposition, natural methods for treatment, and conventional treatments, emphasizing the selection of integrated wastewater treatment, recovery and reuse. This tree includes 10 decision criteria focused on the selection of an optimum treatment alternative. The aim is to identify the lower cost technology, which provides adequate treatment for the local community, corresponding to the available economic resources and trained labour for its sustainable operation and maintenance. Additionally, the method considers possible reuse of the wastewater-treated effluent, and the condition of the receiving water body. Von Sperling (1996) presented a general comparison of aspects to be considered in the selection of wastewater treatment systems from the standpoint of developed countries and for developing countries. This comparison highlights decisive technical criteria in developed countries, such as efficiency, reliability, sludge management, and area requirements. Yang and Kao (1996) developed an expert system for the selection and design of wastewater treatment schemes, considering three critical factors: the type of pollution, the efficiency in removal technologies and the cost of WWTP. Veenstra *et al.* (1997) proposed a selection methodology, considering five general criteria: 1) efficiency of the technology; 2) capability to assimilate water quality and quantity variations; 3) institutional capability to manage the technology; 4) capability to recover operation and maintenance investment costs, including the possibility of reuse, and 5) capability of the technology to comply with the local/national regulatory specifications.

The predictive model. This model was developed by Reid (1982) as a tool to help planners select suitable water supply and wastewater treatment options, which are compatible with available materials and human resource capabilities of a particular local level at a point in time. Several treatment processes are combined and evaluated in relation with the operating constraints, such as limitation of skilled manpower and material requirements. Therefore, a successful selection method also needs to consider socio-economic conditions and local resources.

Analytical Hierarchy Process (AHP). Ellis and Tang (1991) and Tang *et al.* (1997) used AHP in the technology selection of wastewater treatment process, taking advantage of the possibility to include environmental, social, and cultural factors in making decisions. Zeng *et al.* (2007) proposed a multi-criteria analysis methodology including the Analytic Hierarchy Process (AHP) and Grey Relational Analysis (GRA). The process employs a systematic comparison method to select the most appropriate system for the specific user community. In the modelling process, sets of treatment alternatives were formulated in a hierarchical order. The model aims to prioritize a set of weighting variables, such as alternative treatment technology, so that the optimal selection can be made from the priority list of the rankings.

Expert systems, sometimes referred to as *knowledge-based systems*, are computer programs, which provide expert advice, decisions, and a recommended solution for a given situation. They were designed to capture the non-numeric factors and their reasoning logic, which could not be represented in traditional computing approaches, through a set of rules or decision trees (Lukasheh *et al.*, 2001). For the Colombian and Latin-American context, Galvis *et al.* (2005) developed SELTAR for populations under 30,000 inhabitants. This model selects sustainable technologies, considering: characteristics of the technologies, effluent quality, treatment objectives, water uses of the receiving water body, and the initial investment and O & M costs. The model also evaluates the possibility for reuse and considers the socioeconomic and cultural characteristics of the communities. SELTAR considers 104 wastewater treatment schemes and nine technologies for treatment and final disposal of sludge. It has been validated for Colombian municipalities.

Multi Utility Technique (MAUT) is based on a compensatory strategy, called Utility Theory. This involves comparing alternatives that have strengths or weaknesses with regard to multiple objectives of interest to the decision maker. Multi-Attribute Utility Theory (MAUT) is a structured methodology designed to handle the trade-offs among multiple objectives. Utility theory is a systematic approach for quantifying an individual's preferences. It is used to rescale a numerical value on some measure of interest onto a 0-1 scale with 0 representing the worst preference and 1 the best. This allows the direct comparison of many diverse measures. Early applications of MAUT focus on public sector decisions and public policy issues. These decisions not only have multiple objectives; they also often involve multiple constituencies that will be affected in different ways by the decision (Edwards and Newman, 1982). SANEX™ is an example of this type of methodology. This model was developed by Loetscher (1999). It considers community characteristics and is oriented to countries under development and is applicable in locations lacking the infrastructure for disposal and collection of waste, similar to those prevailing in Southern Asia. This model considers the following evaluation criteria: treatment area available, groundwater characteristics, population density, access mode to the zone, water supply, disposal of wastewater and anal cleansing methods. SANEX also considers the possibility of having direct discharge of wastewater, taking into account the assimilation capabilities of the receiving water body.

Methods using Multi-Criteria Decision Analysis (MCDA). Some selection models that incorporate multi-criteria analysis are: support systems for the selection of post-treatment alternatives for anaerobic reactor effluents (PROSAB); Water and Wastewater Treatment Technologies Appropriate for Reuse (WAWTTAR) and Process Selection Model (PROSEL). More recent models, such as the Urban Water Optioneering Tool (UWOT) (Makropoulos *et al.*, 2008), facilitate the selection of combinations of water-saving strategies and technologies and support the delivery of integrated, sustainable water management for new developments. Almeida (1997) and Almeida *et al.* (2001) developed the PROSAB model, considering 9 selection process stages: 1) treatment objectives; 2) wastewater characterisation and definition of the expected treated effluent quality; 3) pre-selection of the technologies and processes forming the integrated treatment and reuse systems; 4) definition of the criteria and the controlling variables of the selection process; 5) comparison of the alternatives; 6) election of an auxiliary method that is useful to contrast and compare with the solution found; 7) prioritisation of alternatives; 8) result analysis, and 9) repetition of the entire selection process with stakeholder participation. WAWTTAR was developed at the University of Humbolt, California, USA. The model considers the availability of technical and human resources for the operation and maintenance and the analysis of financial factors and costs as key criteria (McGahey, 1998). The model uses successive selections, being a first filter for the alternatives considered in the technology packages, considering parameters such as efficiency, costs, wastewater quality, and community characteristics. The second filter evaluates the technological alternatives in terms of the reuse requirements, the public health protection parameters, and the discharge standards.

Methods using matrices. Sobalvarro and Batista (1997) have developed a decision matrix in which the characteristics of 12 technologies were correlated. These characteristics include the area required for the treatment system, odour production, and operational features, including climate, soil characteristics and topography. The limitation of the matrix is that it does not allow other treatment alternatives to be evaluated and it also assigns weights that could vary according to the case study. The Ministerio del Medio Ambiente of Colombia (2002) and Morgan *et al.* (1998) also use the method of weightings for the selection process. By giving a score to each of the parameters evaluated, they are differentiated in terms of the aspects which are considered important to a greater or lesser degree in the technology selection. Noyola *et al.* (2013) developed a matrix method for technology selection of WWTP as a support guide for small and medium-sized cities.

2.6 Water management in the city of the future

2.6.1 Sustainable cities, green cities or eco-cities

Besides the term 'city of the future', different terminologies have been introduced such as 'sustainable city', 'green city', or 'eco-city', which all aim towards the sustainable development of urbanized environments. Green cities, sustainable cities, or eco-cities, are cities designed with consideration for social, economic, and environmental impacts, and have a resilient habitat for existing populations, without compromising the ability of future generations to experience the same. These cities are inhabited by people who are dedicated towards minimization of required inputs of energy, water, food, waste, output of heat, air pollution - CO_2, NO_2, methane, and water pollution (ICLEI - Local Governments for Sustainability USA, 2009). Ideally, a sustainable city creates an enduring way of life across the four domains of ecology, economics, politics and culture (James *et al.*, 2015). However, as a minimum a sustainable city should firstly be able to feed itself with a sustainable reliance on the surrounding countryside. Secondly, it should be able to power itself with renewable sources of energy. The aim is to create the smallest conceivable ecological footprint, while producing the lowest quantity of pollution achievable yet at the same time efficiently using the land; composting used materials, and recycling or converting waste to energy. All of these contributions will lead to the city's overall impacts on climate change to be minimal and with little impact. The challenges of the cities of the future embrace the most vital water needs of urban communities now and in the future, in terms of both water quantity and quality. Section 2.6.2 focuses on the challenges of water management in these cities in comparison with traditional cities.

For today's cities to become sustainable sites, transformational shifts need to be accomplished in three main sectors: energy, water, and food. Prioritising these sectors is strategic as this will catalyse the transformation of many other main components of sustainability such as mitigating climate change, rebalancing elemental cycles, stopping biodiversity loss, and reducing air, soil and water pollution. Transforming these sectors will also provide preconditions for sustainable production and consumption and the protection of ecosystems to become reality. It will also generate the prerequisites to address and resolve the persistent syndromes of poverty and inequality (Gijzen, 2019). The 'city of the future' will require a paradigm shift by building or retrofitting cities in a new way in order to achieve the switch required away from the current unsustainable development path (Table 2.3). Binney *et al.* (2010) present a vision for cities of the future comprising eleven principles arranged under four themes as shown in Figure 2.5.

2.6.3 Blue-Green cities

A Blue-Green City aims to recreate a naturally-oriented water cycle while contributing to the amenities of the city by bringing water management and green infrastructure together. This is achieved by combining and protecting the hydrological and ecological values of the urban landscape while providing resilient and adaptive measures to deal with flood events. Blue-Green Cities generate a multitude of environmental, ecological, socio-cultural and economic benefits. The innovative Blue-Green approach to water management in the city aims to satisfy the demands of urban drainage and planning via coherent and integrated strategies, and places value on the connection and interaction between blue and green assets (Everett *et al.*, 2015). Blue-Green Cities aim to reintroduce the natural water cycle into urban environments and provide effective measures to manage fluvial (river), coastal, and pluvial (urban runoff or surface water) flooding receptors (CRWA, 1998; Ahern, 2013; Everett *et al.*, 2015).

Table 2.3 Comparison between the traditional city and the city of the future

Traditional	City of the future
Drainage: Rapid conveyance of stormwater from premises by underground concrete pipes or culverts, curb and gutter street drainage.	*Storage-oriented*: Keep, store, reuse & infiltrate rainwater locally, extensive use of rain gardens, and drainage mostly on surface.
Wastewater: Conveyance to large downstream treatment plants far from the points of reuse.	*Local reuse*: Treat, reclaim and keep a significant portion of used water locally for reuse in buildings, irrigation and providing ecological low flow to streams. Develop innovative 'water chain' approaches.
Urban habitat infrastructure: No reuse, energy inefficient, excessive use of water.	*Green buildings*: Water-saving plumbing fixtures, energy efficient, larger buildings with green roofs.
Water, stormwater and wastewater infrastructure: Hard structural, independently managed.	*Local cluster decentralized management*: Soft approaches, best management practices as a part of landscape, mimicking nature.
Transportation, roads: Overloaded with vehicular traffic and polluting.	*Emphasis on less polluting fuel*: Bring living closer to cities, good public transport, bike paths, and best management practices to reduce water pollution.
Energy for heating and cooling: Energy brought from large distances, no on-site energy recovery, and high carbon emissions.	*Energy recovery and reduction of use*: Part of heat in wastewater recovered & used locally, biogas from waste, use of geothermal, solar & wind energy.
Overuse of potable water: Drinking water is used for all uses; losses in distribution system.	*Use of treated drinking water*: Water from distant sources should be for potable use only, reuse water more, reduced losses in distribution.
Economies of scale: In treatment cost and delivery is driving the systems – the bigger the better.	*Triple Bottom Line and life cycle assessment*: Of the total economic, social and environmental impact.
Community expectation of water quality: Distorted by hard infrastructure such as buried & fenced off streams for flood and/or effluent conveyance.	*'Stream daylighting' and/or re-naturalization*: Of the water bodies with parks, connecting with built areas enhances the value of surrounding neighborhoods and brings enjoyment.

Source: Adapted from Novotny and Brown (2007)

Blue-Green Cities favour a holistic approach and aim for interdisciplinary cooperation in water management, urban design, and landscape planning. Community understanding, interaction and participation in the development of Blue-Green city designs are actively promoted. Blue-Green Cities typically incorporate sustainable urban drainage systems (SUDS).

2.6.2 Water management in traditional cities versus the city of the future

To ensure the sustainability of the cities of the future it will be necessary to develop innovative ways to consume our limited resources, without diminishing them or degrading the delicate ecological systems on which they depend. Regarding water in the city of the future, we must reform how we manage water resources, water uses, and water infrastructure, so that the water can be re-used several times, and on a city-wide scale via innovative 'water chain' approaches (Gijzen, 2019).

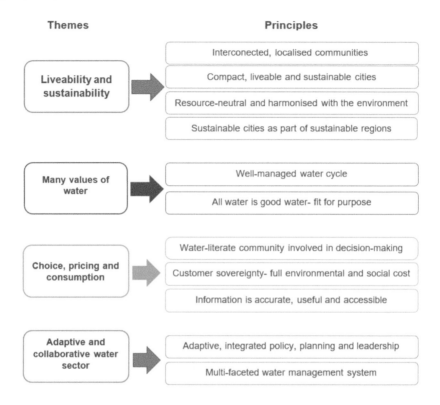

Figure 2.5 Principles for a city of the future
(Binney et al., 2010)

The impacts of climate change, rapid urbanisation and the deterioration of outdated infrastructure, among others, are causing flooding events, water scarcity and rising rehabilitation costs on a scale that will overwhelm the capacities of many cities (Philip *et al.*, 2011). For a sustainable future, sustainable solutions need to be found now so that present issues are resolved without creating new problems for the future (Jefferies and Duffy, 2011). Here are some examples of this new approach:

New-generation systems. Water reuse, rainwater harvesting, grey water recycling; ecosan and urine separation and use; waterless toilets; water-saving devices; natural systems for treatment; soil aquifer treatment and aquifer recharge; sustainable drainage - green/brown roofs, wetlands, ponds, basins, permeable paving; urban agriculture. As previously mentioned, an innovative example is the so-called 'new water' in Singapore, where sewage is treated to generate safe drinking-quality water (Tortajada, 2006; Public Utilities Board, 2016)

'Run to failure'. A concept in asset management where it is more efficient to stop repairing the old systems and eventually replace them with new-generation systems (Nelson, 2008).

Decentralization. Decentralized household and community-scale systems are being widely considered as an alternative response to the deficiencies of centralised approaches in many urban areas as they use fewer resources and are more ecologically benign. The decentralized infrastructure of distributed clusters is the best way to exploit alternative water sources (Wilderer, 2001). In this approach, locally available sources such as rainwater/storm water, local groundwater, and reclaimed wastewater become potential sources of water to offset the freshwater demand from the central water supply system (Libralato *et al.*, 2012; Vairavamoorthy *et al.*, 2015). With decentralization, reuse is facilitated, which implies a reduction in the pressure on water resources (Burkhard *et al.*, 2000; Gijzen, 2006).

Instrumentation, Control and Automation (ICA). ICA is more than Information Technology (IT) or Information and Communications Technology (ICT), but includes all of the following aspects: understanding process dynamics; the development and follow-up of adequate sensors and instrumentation; data handling, telemetry and communication; data and information management; process control and automation; the conversion of data into information for decision making; Edge Processing; dynamic system modelling and simulation in view of design and control (Grievson *et al.*, 2016). ICA can provide the tools for monitoring and controlling urban water management. ICA can be useful for early warning systems related to the impact of pollutant discharges on water bodies (Flores *et al.*, 2014; Velez *et al.*, 2014). Another option is the interaction with satellites to obtain hydro-climatological information, the estimation of precipitation and the temporal and spatial variation of variables of importance for models that calculate both pollutant loads as well as rain-runoff phenomena (Borsanyi *et al.*, 2014; Collins, 2014; Herrero, 2014; Martinez-Cano *et al.*, 2014).

2.6.4 Water Sensitive Urban Design

Cities around the world face a range of critical challenges in managing water resources in terms of quantity and quality. Further, when cities and towns are constructed, the natural landscape is dramatically altered: vegetation and soil are replaced with hard, impervious surfaces and buildings. This leads to the development of unique urban climates that are quite different from those of the surrounding natural environments. This results in increased air pollution, modified rainfall patterns, changed run-off behaviour and higher air temperatures. These challenges have triggered research and developments towards innovative solutions to create more water-sensitive cities and towns (Wong and Brown, 2008 and 2009). This thinking has evolved through the innovation of new concepts such as Water Sensitive Urban Design (WSUD), which is based on the integration of two key fields including 'Integrated urban water cycle planning and management' (IUWCM) and 'urban design' (Figure 2.6). WSUD brings the 'sensitivity for water' to urban design, as it seeks to ensure that water receives due importance within the urban

design process.   WSUD is an interdisciplinary concept of social and physical sciences that
represents context and place (Wong and Ashley, 2006).

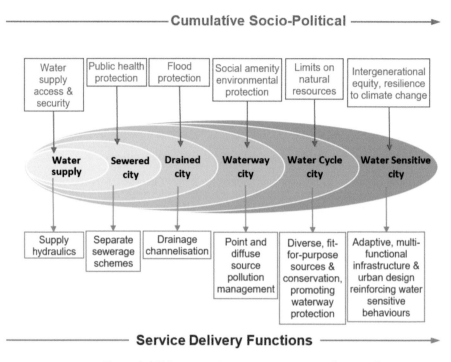

Figure 2.6 Urban water management transitions framework
(Wong and Brown, 2008)

2.6.5 Toward hydrological and ecological sustainability

When looking towards the future of cities, the evolving paradigm is a model of integration of
both new and older urban development including the landscape, drainage, transportation and
habitat infrastructure. This integration will make cities resilient to extreme hydrological events
and pollution, while providing an adequate amount of clean water for sustaining healthy human,
terrestrial and aquatic life. It also supports the creation of an optimal balance between different
economies' uses of water (Novotny and Brown, 2007). Sustainable cities of the future will
combine concepts of 'smart green' development, interconnected ecotones (parks, river riparian
zones), and the control of diffuse and point source pollution from the surrounding areas. They
will be based on reuse of treated effluents and urban stormwater for multiple purposes including
landscape and agricultural irrigation; groundwater recharge to enhance groundwater resources,
environmental flow enhancement of effluent-dominated and flow deprived streams; and
ultimately for water supply.

The paradigm of the city of the future will evolve from the concept of the total hydrologic water and mass balance where all the components of water supply, stormwater, and wastewater will be managed in a closed loop. It will incorporate landscape changes including less imperviousness, more green space used as buffers and for groundwater recharge and it will help restore the landscape's hydrological and ecological functions. It will rely on greatly enhanced removal of organic chemicals, nutrients and endocrine disruptors from effluents and will promote the application of best-management practices providing treatment, water conservation, and storage of excess precipitation for reuse. Closing the water loop may require decentralization of some components of the urban water cycle in contrast to the current highly centralized regional systems employing long-distance water and wastewater transfers (Sitzenfreia and Raucha, 2014). One of the goals of the paradigm is to develop an urban landscape that mimics the natural system that existed before urbanisation. Eco-mimicry includes hydrological mimicry, where urban watershed hydrology imitates the pre-development hydrology, relying on reduction of imperviousness, increased infiltration, surface storage and use of plants that retain water. It will also include interconnected green ecotones around urban water resources that provide habitat to flora and fauna, while providing storage and infiltration of excess flows and buffering pollutant loads from the surrounding urban surfaces.

2.6.6 Reliable, resilient and sustainable water management: the 'Safe & SuRe' approach

To face the challenges of the 21st century a new approach for water management in cities has been proposed under the term 'Safe & SuRe'. This means the design of systems for safe service provision considering the sustainability and resilience to emerging threats. Sustainability and resilience are both dynamic concepts (although over different timescales) that can be incorporated into the water systems not only to avoid negative impacts but also to promote positive ones, yet neither being at the expense of reduced safety (Butler *et al.*, 2014). The Safe & SuRe approach has been developed and designed to demonstrate how emerging threats are able to produce several consequences on society, the economy and the environment. It also clarifies the role of the city water infrastructure in the mediation between threat and impact through compliance with defined levels of service. Key to developing a 'Safe & SuRe' system is to understand which interventions are required. It considers four types of intervention which are presented in Figure 2.7: mitigation, adaptation, coping, and learning (Butler *et al.*, 2016).

Mitigation. Mitigation addresses the link between threat and system and typically denotes long-term actions to ameliorate threats that, although carried out locally, could have wider benefits. In this context, mitigation is defined as 'any physical or non-physical action taken to reduce the frequency, magnitude or duration of a threat'. Reducing greenhouse gas emissions would be an example of a mitigation measure that may be employed both locally and globally to reduce the magnitude of global warming in the long term.

Adaptation. Adaptation measures are interventions that address the link between system and impact and deal with system failures that result from threats that cannot be (immediately) mitigated. Adaptation is typically considered to entail targeted actions or adjustments carried out in a specific system in response to actual or anticipated threats in order to minimize failure consequences.

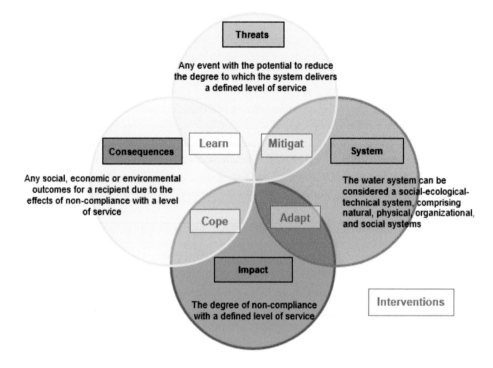

Figure 2.7 Intervention framework of the Safe & SuRe approach
(Butler et al., 2016)

Coping. Within the 'Safe and SuRe' framework, coping addresses the link between impact and consequence. It is defined as 'any preparation or action taken to reduce the frequency, magnitude or duration of the effects of an impact on a recipient'. Coping is often temporary and is actualized should existing mitigation and adaptation measures be insufficient to ensure compliance with required levels of service.

Learning. The final intervention, therefore, is learning, which is placed at the intersection of consequences and threats in the framework and defined as 'embedding experiences and new knowledge in best practice'. There are many approaches to learning, which can include learning from past events, developing pilot schemes to generate new knowledge for best practice, and learning from others. Good data collection and effective communication strategies can also facilitate learning. In all cases, it is important that lessons are learnt from both good and bad practices.

2.7 Water resources management and sustainable development goals

2.7.1 From Millennium Development Goal to Sustainable Development Goals

In September 2015, the United Nations General Assembly unanimously adopted the Sustainable Development Goals (SDGs). The importance of water as an integral part of all human development and ecosystem needs is emphasized through the dedicated Water Goal SDG 6. While many of the Millennium Development Goal (MDG) targets for 2015 have been met or even passed, the MDG target of halving the share of the population without access to basic sanitation was missed by 9 percentage points. However, in absolute numbers, due to population growth, the total number of people without basic sanitation remained almost the same. While major resources have been allocated to health care, education and other development priorities since 2000, the sanitation gap has not been prioritized. Sanitation has therefore been identified as one of 'the most lagging' of all the MDG targets. Furthermore, with their focus on sanitation access and their failure to address wider issues of wastewater and excreta management, the MDGs offered little incentive for investment in more sustainable systems. Thus, much of the sanitation and wastewater management development that has already taken place will require additional investment to make it both more effective and more sustainable. The universal applicability and emphasis on integrated solutions in the SDGs and the broader 2030 Agenda provide strong arguments for investing in sustainable sanitation and wastewater management. The SDGs dedicate an entire goal to water and sanitation via SDG 6 'to ensure availability and sustainable management of water and sanitation for all', bringing greater awareness to sanitation challenges (Andersson, 2016).

2.7.2 Sustainable Development Goal 6: Ensure access to water and sanitation for all

SDG 6 has two targets which are directly linked to sanitation and wastewater management: Target 6. 2.... '.... achieve access to adequate and equitable sanitation and hygiene for all, and end open defecation, paying special attention to the needs of women and girls and those in vulnerable situations'; Target 6.3:.. '.... improve water quality by reducing pollution, eliminating dumping and minimizing release of hazardous chemicals and materials, halving the proportion of untreated wastewater, and substantially increasing recycling and safe reuse globally'. Goal 6 goes beyond drinking water, sanitation and hygiene to also address the quality and sustainability of water resources. Besides, it is recognised that the success of all the other SDGs will heavily depend on water. Agenda 2030 recognizes the centrality of water resources for sustainable development and the vital role that improved drinking water, sanitation and hygiene play in progress in other areas, including health, food security, education, sustainable cities, and poverty reduction (United Nations, 2016). Sustainable water and sanitation and wastewater management is influential within all the SDGs.

Sustainable sanitation (SDG 6) can also make cost-effective contributions to achieving a wide variety of other SDG goals and targets (Hall *et al.*, 2016). The number of targets addressed can increase with the level of ambition in sustainable sanitation and wastewater management investments. For example, at the most basic levels of ambition (ending open defecation and preventing human exposure to pathogens and toxic substances in excreta and wastewater), improving sanitation and wastewater management could relieve a large burden of infectious disease (Goal 3), particularly child mortality. Lower incidence of disease means fewer days of education (Goal 4) and of productive work lost (Bos *et al.*, 2004). If systems also aim to prevent the release of untreated wastewater in natural ecosystems and to reduce the run-off of nutrients from agricultural soil caused by fertilizer application, they could improve the status of freshwater and coastal ecosystems and the services they provide (Goal 14). Recovering and reusing the valuable resources present in excreta and wastewater also contributes to resource efficiency (Goal 12), conservation of freshwater ecosystems and restoring degraded land and soils (Goal 15) (Jenkins, 2016; WHO, 2016), and can help improve food security (Goal 2). Sustainable sanitation and wastewater management value chains provide new livelihood opportunities (Goals 1 and 8). To make tomorrow's cities liveable (Goal 11) it is necessary to introduce adequate sanitation and wastewater management. Furthermore, 'equitable accesses to adequate sanitation can also help to achieve non-discrimination targets under Goal 5 by increasing equal participation in school, the workforce, institutions and public life. A lack of suitable facilities excludes women, girls and people with disabilities and increases the risk of gender-based violence (Andersson *et al.*, 2016). Other goals such as Goal 7 on renewables and energy efficiency will reinforce targets related to water pollution and aquatic ecosystems by reducing levels of chemical and thermal pollution (compared to a less efficient fossil energy supply system). Climate change (Goal 13) will manifest mainly by sea level rise and the intensification of the hydrological cycle, producing more frequent and intense rainfall as well as extended dry periods. As a result, a city's water supply, wastewater and stormwater systems will be particularly affected. Constructing new greener infrastructures, retrofitting or reconfiguring existing infrastructure systems and exploiting the potential of smart technologies (Goal 9) can greatly contribute to the reduction of environmental impacts and disaster risks as well as the construction of resilience and the increase of efficiency in the use of water resources (GWSP, 2015).

2.7.3 Governance processes for the Sustainable Development Goals

The implementation of the ambitious SDGs poses considerable challenges for water governance. Many problems related to water arise from inadequate and dysfunctional governance settings, irrespective of whether water scarcity is prevalent or not. A lack of institutional capability is a central factor to explain the poor performance of water governance in many countries. Effective implementation of the SDGs requires adaptive and effective governance and the adherence to good governance principles in water-related sectors and elsewhere to prevent adverse implications. The SDG implementation process must thus support the building of institutional capacity to achieve its goals (GWSP, 2015; Hall *et al.,* 2016).

2.8 Research gaps and research questions

2.8.1 Main conclusions form the literature review

The literature review demonstrates that:

- The exponential growth rate of the human population, as well as agricultural and industrial expansion, have generated an increase in freshwater supply demand, and consequent challenges of access. Currently, in many parts of the world, there is a (looming) crisis due to challenges arising from the low quantity of available water (water scarcity by quantity).

- In most cases, used water is returned to water resources as untreated wastewater, leading to water quality deterioration, which negatively impacts on aquatic habitats and the quality of life of communities, with subsequent economic, social and environmental impacts. Globally, approximately 80% of wastewater is released into the environment without treatment.

- Unlike point source pollution, which enters a river course at a specific site, usually via pipe discharge, diffuse pollution (rural, agricultural and urban) occurs when polluting substances leach into surface waters and groundwater as a result of rainfall, soil infiltration or surface runoff. In developing countries, this type of pollution has not yet received much attention, while in developing countries, regulations and investments to reduce their impacts have had limited success.

- In addition to the classical parameters of contamination by organic matter and pathogens associated with point-source pollution by domestic wastewater, in the last decades there has been concern about the contamination of water by micro-pollutants. Higher concentrations of persistent organic pollutants (point and diffuse sources) have been found in food chains exposing humans and wildlife to toxic effects of this kind of pollution.

- Traditionally, the assessment of water scarcity has primarily focused on water quantity. However, with the increasing deterioration of water quality, in many cases water quality does not meet the minimum quality requirements that its different uses require (water scarcity by quality).

- The continuing global increase in water demand, combined with the escalating freshwater quantity and water quality crisis, presents a phenomenal challenge for many countries in the coming decades to ensure water and food security for their growing populations.

- Climate change makes the problems related to freshwater more critical. Climate change is associated with the intensification of the hydrological cycle, producing more frequent and intense rainfall as well as extended dry periods. As a result, a city's water supply, wastewater and stormwater systems will be particularly affected. Lakes and reservoirs are affected mainly due to variations in water temperature, affecting oxygen regimes, oxidation/reduction reactions, stratification, mixing rates, and the development of biota. The increasing temperature leads to decreases in the self-purification capacity of water bodies by reducing the amount of dissolved oxygen.

- Climate-induced extreme weather events, expressed in longer drought and heavier rainfall periods, will also cause sharper peaks of water flows and run-off pollution in cities, which are likely to surpass the design capacities of a city's water system.

- Water resources are substantially affected by human activities such as dam building, deforestation, mining activities, land use changes and pollutant loads. Other impacts are associated with the building of housing in sensitive areas, such as on high slopes in the upper parts of water catchment areas, and very close to sensitive groundwater aquifers. The erosion associated with deforestation has altered the water cycle and has caused the loss of soil, increasing the sediment load transported towards sewage systems, streams, rivers and coasts.

- Many countries have been adopting pollution control approaches which are based exclusively on 'end-of-pipe' solutions by constructing WWTPs. These approaches for effluent management are ineffective, unsustainable, and costly, and they lead to significant problems related to public health, water quality and the environment, and as such to the economy at large.

- Over the last few decades, a number of new concepts and approaches related to sustainable water management have emerged. Among them, the following stand out: (Water) Resilience Building, Integrated Water Resources Management (IWRM), Hydrological cycle, Urban Water Cycle (UWC), Integrated Urban Water Management (IUWM), Ecohydrology, and Water Governance. Based on these broader concepts, several strategies related to the sustainability of cities and sustainable water management in cities have been developed: Sustainable cities, green cities or eco-cities, Blue-Green cities and Water Sensitive Urban Design (WSUD). The Sustainable Water Management Improves Tomorrows Cities Health (SWITCH) project was a research partnership funded by the European Commission over the period 2006 to 2011. It involved an implementing consortium of 33 partners from 15 countries. SWITCH involved innovation in the area of sustainable urban water management (IUWM). This project looked towards water management in the 'city of the future' and aimed to challenge existing patterns and to find and promote more sustainable alternatives to the conventional ways of managing urban water.

- In order for investments in water and sanitation to produce the expected outcomes (Sustainable Development Goals) and to contribute to quality of life improvement in communities, improved water management, sustainable water supply and sanitation systems and innovative strategies and approaches are required to turn the tide. Criteria and conceptual models have been developed that can help to respond to this challenge.

- By applying the principles of cleaner production, the 3-Step Strategic Approach was proposed as an innovative and integrated way of achieving sustainable management of urban water, nutrients and waste. There is a need to advance in the validation and implementation of innovative models such as the 3-SSA on a real scale.

2.8.2 Research gaps

The literature review also reveals a number of research gaps, including the following:

- Based on the different sustainable water management concepts, different methods and innovative approaches have been proposed such as alternatives to the business-as-usual practise of end-of-pipe solutions. These methods, such as the Three-Step Strategic Approach (3-SSA), need to be reviewed and validated to stimulate and facilitate their implementation in practice.

- Detailed review and comparison of the conventional strategy (end-of-pipe approaches) and innovative strategies, using different methodologies (multi-criteria analysis and CBA) and different criteria (social, technical, environmental, and economic).

- Detailed review of the potential to implement strategies of efficient use of water and minimization and reduction of waste at the household level. This review includes the selection of technology for different alternatives (e.g. change of habits, reuse of grey water, rainwater harvesting) and comparison (CBA) with the conventional alternative, which considers a 'business-as-usual scenario' of high water use, end-of-pipe wastewater treatment plants and the conventional water supply system with drinking water quality for all uses.

- Development and application of methodologies to select urban drainage system technology with the purpose of optimizing the investments considering prevention and minimization of waste production (at the urban water cycle), and the impacts of both point-source and diffuse pollution in the receiving water bodies.

- Development and application of methodologies to make a detailed evaluation of the potential of reuse with treated wastewater, especially in agricultural irrigation, through case studies that consider variables such as: flow, rainfall temporal variation, availability of irrigation area, regulations, WWTP, costs (initial investment and O&M), water tariffs and taxes for wastewater discharges to water bodies.

- Development and application of methodologies to study the effect on water quality and the self-purification capacity of water bodies in scenarios such as: 1) the impact of multi-purpose reservoirs (power generation, flood control and pollution control); 2) the spatial and temporal distribution of pollution, considering the basin as the unit of analysis; 3) the impact of pollution peaks due to diffuse contamination from urban areas (stormwater).

- Considering the basin as a unit of analysis, and development of case studies to compare two overall scenarios for improving urban water management and water resource quality improvement: i) conventional strategy, which considers a 'business-as-usual scenario' of high water use, end-of-pipe wastewater treatment and conventional water supply providing drinking water quality for all uses; and ii) the systematic and chronological implementation of the 3-SSA: 1) prevention or minimisation of water use and waste production; 2) treatment, recovery and reuse of water and waste components, and 3) disposal of water and waste with stimulation of natural self-purification in the receiving water body.

2.8.3 Main topics of literature review addressed in this PhD research

The overall objective of this research thesis is to identify and validate the 3-SSA (Section 2.4.1). The research topics studied in more detail are related to each of the three steps: Step 1: minimization and prevention (sections 2.4.3 and 2.3.6); Step 2: treatment, reuse and recovery of components (Section 2.4.4) and Step 3: disposal of remaining wastewater with stimulation of natural self-purification (Section 2.4.5). The research includes the comparison of results of the application of the conventional strategy (end-of-pipe solutions, Section 2.2.8) versus the systematic application of the three steps (Un-conventional strategy: 3-SSA, Section 2.4.1). The two strategies differ in the application of Integrated Water Resource Management (Section 2.3.3). The basin is the unit of analysis for both strategies, but for un-conventional strategy the investments were prioritized based on water quality objectives (Section 2.1.3).

This PhD research suggests the need to conceive the sewerage, the WWTP and the receiving water body as an integrated system. The technology selection on minimization, prevention and control of both point and diffuse pollution should be considered in this integrated system (sections 2.2.4 and 2.2.5). In this research, to study the feasibility of the three steps, each individually and combined, the following methodological tools were used: CBA, AHP, MCDA and river quality modeling (Section 2.5.1).

2.8.4 Further recommended research

The literature review identifies a number of research topics which are not addressed in this PhD thesis, but which would warrant further attention. This includes for instance further research on: 1) *Water quality indicators*. In this study dissolved oxygen (DO) and biochemical oxygen demand (BOD$_5$) were used as classic indicators of water pollution. However, it is recommended for future research to also include other compounds and indicators such as pesticides, fertilizers, heavy metals, micro pollutants, etc., which may have other (eco-toxicological) impacts, beyond oxygen consumption (Section 2.2.4). For these contaminants the best management options are provided under Step 1 of the 3-SSA (minimisation and prevention); 2) The implementation of cleaner production to minimize and prevent waste flows other than domestic sources (industrial, agricultural); 3) Evaluation of different options under Step 3 *Stimulated natural self-purification, such as using* ecohydrology approaches (Section 2.3.7). This could be addressed via a study on the effect of hydraulic and ecohydrology interventions in the Sonso Lagoon and the effect of floodplains; 4) Evaluate strategies included in the paradigm shift for the city of the future (Section 2.6.2), such as: decentralization, Instrumentation Control and Automation (ICA) and water sensitivity Urban Design WSUD (Section 2.6.4).

2.9 References

Ahern, J. (2013). Urban landscape sustainability and resilience: The promise and challenges of integrating ecology with urban planning and design. *Landsc. Ecol.* 28, 1203-1212.

Almeida, M.A. (1997). Metodología de Análisis de Decisiones para Seleccionar Alternativas de Tratamiento y Uso de Aguas Residuales. Universidad de Brasilia. (In Spanish).

Almeida, M.A., Cordeiro, O. and Lopes, R.P. (2001). Sistema de Apoio Â Desisao para Seleçao de Alternativas de Postratamento de Efluentes de Reatores Anaeróbios. In Chernicharo, C.A.L (ed.) Pós-Tratamento de Efluentes de Reatores Anaeróbios. Finep. Belo Horizonte, Brazil. (In Portuguese).

Anderson, J. and Iyaduri, R. (2003). Integrated urban water planning: big picture planning is good for wallet and the environment. *Water Sci Technol* 47(7–8):19–23, IWA Publishing.

Andersson, K., Rosemarin, A., Lamizana, B., Kvarnström, E., McConville, J., Seidu, R., Dickin, S. and Trimmer, C. (2016). *Sanitation, Wastewater Management and Sustainability: from Waste Disposal to Resource Recovery*. Nairobi and Stockholm: United Nations Environment Programme and Stockholm Environment Institute.

Arboleda, J. (2000). Cronología del desarrollo de la tecnología de la purificación del agua. (In Spanish).

ASCE (American Society of Civil Engineers) (1999). Task Committee on Sustainability Criteria, UNESCO-IHP IV Project M-4.3. *Sustainability Criteria for Water Resource Systems*. Reston, Va., ASCE Publications.

Bates, B.C., Kundzewicz, Z.W., Wu, S. and Palutikof, J.P. (Eds.) (2008). Climate change and water. Technical paper of the Intergovernmental Panel on Climate Change, IPCC Secretariat, Geneva, 210 pp.

Binney, P., Donald, A., Elmer, V., Ewert, J., Phillis, O., Skinner, R. and Young, R. (2010). IWA Cities of the Future Program, Spatial Planning and Institutional Reform Conclusions from the World Water Congress, Montreal.

Bixio, D. and Wintgens, T. (2006). Water Reuse System Management Manual AQUAREC. Office for Official Publications of the European Communities, Luxemburg. ISBN 92-79-01934-1.

Blanco, S.A., Torres, P. and Galvis, A. (2017). Identification of resilience factors, variables and indicators for sustainable management of urban drainage systems. *DYNA* 84(203), pp. 126-133.

Boraschi, S.F. (2009). Corredores biológicos: una estrategia de conservación en el manejo de cuencas hidrográficas. *Kurú: Revista Forestal.* Costa Rica. 6 (17). (In Spanish).

Borsanyi, P., Hamududu, B., Navaratnam, S. and Langsholt, E. (2014). Improvement Of The National Flood Early Warning System In Norway - Flood Level Warnings And Uncertainties. In: 11th International Conference on Hydrionformatics 2014, New York, USA.

Bos, J.J., Gijzen, H.J., Hilderink, H.B.M., Moussa, M., Niessen, L.W. and de Ruyter-van Steveninck, E.D. (2004). Quick Scan Health Benefits and Costs of Water Supply and Sanitation. Netherlands Environmental Assessment Agency (RIVM), Institute for Water Education - (UNESCO-IHE), Erasmus University Rotterdam. Bilthoven, the Netherlands.

Bravo-Inclán, L., Saldaña-Fabela, P., Dávila, J. and Carro, M. (2013). La importancia de la contaminación difusa en México y en el mundo. Technical Report. 10.13140/2.1.3336.7843. (In Spanish).

Bregulla, J., Powell, J., and Yu, C. (2010). A simple guide to Sustainable Drainage Systems for housing. Published by IHS BRE Press on behalf of the NHBC Foundation.

Brent, R. (2006). Applied cost-benefit analysis, Second Edition Ed., Edward Elgar Publishing limited, Cheltenham, UK.

Burian, S.J., Nix, S.J., Pitt, R.E. and Durrans, S.R. (2000). Urban Wastewater Management in the United States: Past, Present, and Future. *Urban Technology* 7(3): 33-62.

Burian, S.J. and Edwards, F.G. (2002). Historical Perspectives of Urban Drainage. In: 9th International Conference on Urban Drainage. *American Society of Civil Engineers* (ASCE). Portland, Oregon.

Burkhard, R., Deletic, A., and Craig, A. (2000). Techniques for water and wastewater management: a review of techniques and their integration in planning. *Urban Water,* 2, 197-221.

Butler, D., Farmani, R., Fu, G., Ward, S., Diao, K. and Astaraie-Imani, M. (2014). A new approach to urban water management: Safe and sure. *Procedia Eng.* 89, 347-354.

Butler, D., Ward, S., Sweetapple, C., Astaraie-Imani, M., Diao, K., Farmani, R. and Fu, G. (2016). Reliable, Resilient and Sustainable Water Management: The Safe & SuRe Approach. *Glob. Challenges*, 1-15.

Cardona, M.M. (2007). Minimización de Residuos: una política de gestión ambiental empresarial. *Producción más Limpia* 1(2), 46-57. (In Spanish).

Castro, S.I. (2008). La cuenca urbana en la ciudad media : relaciones de conflicto entre ecosistema y ciudad. (In Spanish). [Online] URL: *caleidoscopiosurbanos.com/.../images_caleidoscopiosurbanos_articulos-y-publicacion.*

Chanan, A.P., Vigneswaran, S., Kandasamy, J. and Bruce, S. (2013). Wastewater Management Journey – From Indus Valley Civilisation to the Twenty-First Century. In: Sharma, S., Sanghi, R. (Eds.), Wastewater Reuse and Management. Springer, London, UK, pp. 3-17.

CIRIA (2015). The SUDS Manual. London.

Collins, F. (2014). High-resolution flash flood forecasting for large urban areas – sensitivity to scale of precipitation. In: 11th International Conference on Hydrionformatics 2014, New York, USA.

Cooper, P.F. (2001). Historical Aspects of Wastewater Treatment. In: *Decentralised Sanitation and Reuse: Concepts, Systems and Implementation.* In Lens, P.G. Zeeman, and G. Lettinga (Eds.). IWA Publishing.

Corcoran, E., Nellemann, C., Baker, E., Bos, R., Osborn, D. and Savelli, H. (2010). Sick water?: the central role of wastewater management in sustainable development: a rapid response assessment, GRID Arendal Ed., UNEP/Earthprint, Norway.

Crites, R. and Tchobanoglous, G. (1998). Small and Decentralized Wastewater Management Systems. Mcgraw-Hill. USA.

Crittenden, J.C. (2015). Water for Everything and the Transformative Technologies to Improve Water Sustainability. National Water Research Institute. The 2015 Clarke Prize Lecture.

Cruz, J.R. (2015). Manejo eficiente del riego en el cultivo de la caña de azúcar en el valle geográfico del río Cauca. Cali. Centro de Investigaciones de la Caña de Azúcar. Cenicaña. Cali. (In Spanish).

CRWA (1998). Blue Cities Guide: Environmentally Sensitive, Blue Cities Initiative. Boston, USA.

De Toffol, S., Engelhard, C. and Rauch, W. (2007). Combined sewer system versus separate system – a comparison of ecological and economic performance indicators. *Water Science & Technology,* 55 (4), 255-264.

Dunning, D.J., Ross, Q.E. and Merkhofer, M.W. (2000). Multiattribute utility analysis; best technology available; adverse environmental impact; Clean Water Act; Section 316(b). *Environ Science Policy* 3:7–14.

Edwards, W. and Newman, J.R. (1982). Quantitative Applications in the Social Sciences: Multiattribute evaluation Thousand Oaks, CA: SAGE Publications Ltd doi: 10.4135/9781412985123

Ellis, K.V. and Tang, S.L. (1991). Wastewater Treatment Optimization Model for Developing Worlds. I: Model Development. *Journal of Environmental Engineering.* 117(4): 501-518.

Enderlein, U.S., Enderlein, R.E. and Willimas, P.W. (1997). Water Quality Requirements. Water Pollution Control - A guide to the use of water quality management principles. In Helmer, R. and Hespanhol, I. (Eds.). Published on behalf of WHO/UNEP by F & FN Spon, London, UK.

Environment Agency (2007). The unseen threat to water quality. Diffuse water pollution in England and Wales report, May 2007.

Erickson, B.E. (2002). Analysing the ignored environmental contaminants. *Environ Science and Technology*: 36 (7): 140–145.

European Commission - DG environment (2008). Water Note 1. Joining Forces for Europe's Shared Waters: Coordination in International River Basin Districts. Water Information System for Europe (WISE). Kongens Nytorv, Denmark.

Everett, G., Lamond, A. and Lawson, A. (2015). Green Infrastructure. In: Sinnett, D., Burgess, S., and Smith, N. (Eds.) Handbook on Green Infrastructure: *Planning, Design and Implementation.* Cheltenham, Bristol, UK, pp. 50-66.

Fletcher, T.D., Shuster, W., Hunt, W.F., Ashley, R., Butler, D., Arthur, S., Trowsdale, S., Barraud, S., Semadeni-Davies, A., Bertrand-Krajewski, J.L., Mikkelsen, P.S., Rivard, G., Uhl, M., Dagenais, D. and Viklander, M. (2015). SUDS, LID, BMPs, WSUD and more – The evolution and application of terminology surrounding urban drainage. *Urban Water,* 12, 1-18.

Flores, E., Thouvenel, F., Mazella, J., Thouvenel, T. and Soyeux, E. (2014). Ripost : a river pollution simulation tool to anticipate the consequence of accidental spills. In: 11th International Conference on Hydroinformatics 2014, New York, USA.

Folke, C. (2006). Resilience: The emergence of a perspective for social-ecological systems analyses. *Global Environmental Change-Human and Policy Dimensions,* 16, 253-267.

Fürhacker, M., McArdell, S.M, Lee, Y., Siegrist, H., Ternes, T.A., Li, W. and Hu, J. (2016). Assessment and Control of Hazardous Substances in Water. In Global Trends & Challenges in Water Science, Research and Management. A compendium of hot topics and features from IWA Specialist Groups Second Edition. International Water Association (IWA) Alliance House, London, UK. ISBN 9781780408378.

Galvis, A., Cardona, D.A. and Bernal, D.P. (2005). Modelo Conceptual de Selección de Tecnología para el Control de Contaminación por Aguas Residuales Domésticas en Localidades Colombianas Menores de 3000 Habitantes SELTAR. Proceedings IWA Agua 2005. Conferencia Internacional: De la Acción Local a las Metas Globales, November, 2005, Cali, Colombia. (In Spanish).

Galvis, A., Guerrero, J.E., Saldarriaga, G. and Buitrago, C.A. (2006). Proyecto Selección de Tecnología para el Control de Contaminación por Aguas Residuales Domésticas. *Acodal*, 49 (214). (In Spanish).

Galvis, A., Zambrano, D., van der Steen, N.P. and Gijzen, H.J. (2014). Evaluation of a pollution prevention approach in the municipal water cycle, *Cleaner Production*. 66, 599-609.

Garcia-Quiroga and Abad-Soria (2014). Los corredores ecológicos y su importancia ambiental: Propuestas de actuación para fomentar la permeabilidad y conectividad aplicadas al entorno del río Cardeña (Avila y Segovia). *Observatorio Medioambiental*, 17, 253-298. (In Spanish).

Gijzen, H.J. (1999). Sustainable wastewater management via re-use: Turning waste into wealth. In: Garcia, M. et al (Eds.) Proceedings AGUA 98 - Water and Sustainability, Cali, June 1-3, 1998, pp 211-224.

Gijzen, H.J. (2001a). Low Cost Wastewater Treatment and Potentials for Reuse A cleaner production approach to wastewater management. International Symposium on Low-Cost Wastewater Treatment and Re-use, NVAWUR-EU-IHE, February 3-4, 2001, Cairo, Egypt.

Gijzen, H.J. (2001b). Anaerobes, aerobes and phototrophs. A winning team for wastewater management. *Water Science and Technology* 44(8), 123-132.

Gijzen, H.J. (2006). The role of natural systems in urban water management in the City of the Future - A 3-step strategic approach. *Ecohydrol. Hydrobiol. 6, 115-122.*

Gijzen H.J. (2019 - in press). The SDGs in the Urban Context – an opportunity to shape sustainable cities. In (G. El-Khoury et al, eds): Inclusive Cities for 2030: Towards an enabling environment for urban inclusion - An edited volume.

Gleick, P.H. (2009). Doing more with less: improving water use efficiency nationwide. Southwest Hydrologic.

González, S., Lopez-Roldan, R. and Cortina, J.L. (2012). Presence and biological effects of emerging contaminants in Llobregat river basin: A review. *Environmental Pollution*, 161(2012), pp 83-92.

Grievson, O., Baeza, J. A., Thompson, K., Ingildsen, P., Olsson, G. and Volcker, E. (2016). Instrumentation, Control and Automation. In Global Trends & Challenges in Water Science, Research and Management. A compendium of hot topics and features from IWA Specialist Groups Second Edition. International Water Association (IWA) Alliance House, London, UK. ISBN 9781780408378.

Grigg, N.S. (2016). IWRM and Water Governance. In: Integrated Water Resource Management. Palgrave Macmillan, London. DOI https://doi.org/10.1057/978-1-137-57615-6_5.

GWP (1992). Dublin Principles. Stockholm, GWP. http://www.gwp.org/The-Challenge/What-is-IWRM/Dublin-Rio-Principles/

GWP (2012). Integrated Urban Water Management, TEC Background Papers. Stockholm, Sweden.

GWP and INBO (2009). A Handbook for Integrated Water Resources management in Basins. ISBN: 978-91-85321-72-8.

GWP and SAMTAC (2000). Agua para el Siglo XXI: de la Visión a la Acción. Report produced by South American Technical Advisory Committee (SAMTAC), *Global Water Partnership (GWP)*. Buenos Aires, Argentina, pp 81.

GWSP (2015). Towards a Sustainable Water Future- Sustainable Development Goals: A Water Perspective.
Fileadmin/images/SDG_CONF/Towards_a_Sustainable_Water_Future.

Hajkowicz, S. and Collins, K. (2007). A review of multiple criteria analysis for water resource planning and management. *Water Resources Management*, 21, 1553-1566.

Hajkowicz, S. and Higgins, A. (2008). A comparison of multiple criteria analysis techniques for water resources management. *Operational Research* 184, 255-265.

Hall, N., Richards, R., Barrington, D., Ross, H., Reid, S., Head, B., Jagals, P., Dean, A., Hussey, K., Abal, E., Ali, S., Boully, L. and Willis, J. (2016). Achieving the UN Sustainable Development Goals for water and beyond, Global Change Institute, The University of Queensland, Brisbane.

Hamouda, M.A. (2011). Selecting Sustainable Point-of-Use and Point-of-Entry Drinking Water Treatment: A Decision Support System. PhD Thesis, Civil Engineering. University of Waterloo, Ontario, Canada.

Harremoes, P. (2000). Advanced water treatment as a tool in water scarcity management. *Water Science and Technology*, 42 (12), 73-92.

Harrison, M. (2010). Valuing the Future: the social discount rate in cost-benefit analysis, Visiting Researcher Paper, Productivity Commission, Canberra.

Heinz, I., Pulido-Velazquez, M. and Lund, J.R. (2007). Hydro-economic Modelling in River Basin Management: Implications and Applications for the European Water Framework Directive Water Resource Management 21: 1103. https://doi.org/10.1007/s11269-006-9101-8.

Herrero, J. (2014). Flood Alert System for Early Warning in Mountainous Coastal Watersheds : Coupling Data-Driven and Physically Based Hydrological Models. In: 11th International Conference on Hydrionformatics 2014, New York, USA.

Hespanhol, I. (2003). Potencial de reúso de água no Brasil. In: Reúso de água. Chapter 3, P.C.S. Mancuso and H.F. d. Santos, Eds., Universidade de Sáo Paulo Faculdade de Saúde Pública. ABES.

Holling, C.S. (1973). Resilience and stability of ecological systems. *Annual Review of Ecology Evolution and Systematics*, 4, pp. 1-23.
DOI: 10.1146/annurev.es.04.110173.000245 99(HY10), pp. 1837-1848.

Hynes, H.B.N. (1960). The biology of polluted waters, 1st edition, Liverpool University Press.

ICLEI - Local Governments for Sustainability USA (2009). Sustainability planning toolkit. A comprehensive guide to help cities and counties develop a sustainability plan. https://www.hud.gov/sites/documents/20399_iclei_sustainabil.pdf

James, P., Magee, L., Scerri, A. and Steger, M.B. (2015). Urban Sustainability in Theory and Practice: Circles of Sustainability. London: Routledge.

Jefferies, C. and Duffy, A. (2011). SWITCH Transition Manual. SWITCH Document. Deliverable reference: D1.3.4.

Jenkins, M. (2016). Access to Water and Sanitation. Parliamentary Office for Science and Technology (POST) note (Vol. 521). London.

Juuti, P., Katko, T. and Vuorinen, H. (2008). Environmental History of Water - Global views on community water supply and sanitation. *Water Intelligence Online* Vol. 6, 2007, DOI: 10.2166/9781780402277

Kundzewicz, Z. W., Mata, L.J., Arnell, N.W., Doll, P., Kabat, P, Jimenez, B, Miller, KA, Oki, T, Sen, Z. and Shiklomanov, I.A. (2007). Freshwater resources and their management. In: Parry, M.L., Canziani, O.F., Palutikof, J.P., van der Linden, P.J., Hanson, C.E. (Eds.) Climate change 2007: impacts, adaptation and vulnerability. Contribution of Working Group II to the fourth assessment report of the Intergovernmental Panel on Climate Change. Cambridge University Press, Cambridge, pp 173–210.

Larsen, H., Ipsen, N.H. and Ulmgren, L. (1997). Policy and Principles. Water Pollution Control - A guide to the use of water quality management principles. In Helmer, R. and Hespanhol, I. (Eds.). Published on behalf of WHO/UNEP by F & FN Spon, London, UK.

Libralato, G., Volpi Ghirardini, A. and Avezzù, F. (2012). To centralise or to decentralise: An overview of the most recent trends in wastewater treatment management. *Environmental Management,*. 94, pp. 62-68.

Liou, S.M., Lo, S.L. and Hu, C.Y. (2003). Application of two-stage fuzzy set theory to river quality evaluation in Taiwan, *Water Research,* 37(6), pp. 1406-1416.

Liu, S., Butler, D., Memon, F. A., Makropoulos, C., Avery, L., and Jefferson, B. (2010). Impacts of residence time during storage on potential of water saving for grey water recycling system. *Water Research*, 44(1), pp. 267-277.

Loetscher, T. (1999). Appropriate Sanitation in Developing Countries: The Development of a Computerized Decision Aid. Ph.D. Dissertation. University of Queensland. Brisbane, Australia.

Lofrano, G. and Brown, J. (2010). Wastewater management through the ages: A history of mankind. *Science of the Total Environment*, 408, 5254-5264.

Loftus, A.C. (2011). Adapting urban water systems to climate change. A handbook for decision makers at the local level. ICLEI. SWITCH Project, supported by the European Commission under the 6th Framework Programme Contract No. 018530-2. ISBN 978-3-943107-10-4.

Lukasheh, A.F., Droste, R.L. and Warith, M.A. (2001). Review of Expert System (ES), Geographic Information System (GIS), Decision Support System (DSS), and Their Applications in Landfill Design and Management. *Waste Management and Research.* 19: 177-185.

Makropoulos, C.K., Natsis, K., Liu, S., Mittas, K. and Butler, D. (2008). Decision support sustainable option selection in integrated urban water management. *Environmental Modelling & Software*, 23 (12), pp. 1448-1460.

Marsalek, J., Jimenez-Cisneros, B., Karamouz, M., Malmquist, P.M., Goldenfum, J., Chocat, B., (2008) Urban water cycle processes and interactions. *Urban Water Series –* UNESCO – PHI. ISBN 978-0-415-45347-9.

Martinez-Cano, C., Galvis, A., Alvis, F. and Werner, M. (2014). Model integration to improve an early warning system for pollution control of the Cauca River. In: 11th International Conference on Hydroinformatics 2014, New York, USA.

McClain, M.E. (2008). Ecohydrology as a tool in the sustainable development of large tropical rivers. In: Harper, D, Zalewski, M., Pacini, N. (Eds.) Ecohydrology: processes, models and case studies. CABI, Oxfordshire.

McGahey, C. (1998). Water and Wastewater Treatment Technologies Appropriate for Reuse WAWTTAR. CEP Technical Report #43, Caribbean Environment Programme, United Nations Environment Programme.

Mejía, F.J., Isaza, P.A., Aguirre, S. and Saldarriaga, C.A. (2004). Reutilización de aguas domésticas. In: XVI Seminario Nacional de Hidráulica e Hidrología, Armenia, Colombia. (In Spanish).

Metcalf and Eddy, Inc. (1991). Wastewater Engineering. Treatment, Disposal, Reuse. McGraw Hill International Edition. Singapore.

Metcalf and Eddy, Inc. (1995). Ingeniería de Aguas Residuales. Tratamiento, Vertido y Reutilización. McGraw Hill. Barcelona, Spain. (In Spanish).

Metcalf and Eddy, Inc. (2003). Wastewater Engineering. Treatment and Reuse. McGraw Hill. New York, USA. Fourth Edition.

Ministerio del Medio Ambiente de Colombia (2002). Sistemas de Alcantarillado y Plantas de Tratamiento de Aguas Residuales. Guía Ambiental. Bogotá, Colombia. (In Spanish).

Miranda, J. (2000). Gestión de proyectos: identificación, formulación, evaluación financiera, económica, social, ambiental, M.M (Ed.). (In Spanish).

Mitchell, V.G. (2006). Applying integrated urban water management concepts: A review of Australian experience. *Environmental Management*. 37, pp. 589-605.

Moench, M., Dixit, A., Janakarajan, M., Rathore, S. and Mudrakartha, S. (2003). The fluid mosaic, water governance in the context of variability, uncertainty and change. Nepal Water Conservation Foundation, Kathmandu, and the Institute for Social and Environmental Transition, Boulder, Colorado, USA.

Monerris, M., and Marzal, P. (2001). Modelación de la Calidad del Agua. U.P.V. (Ed.) Valencia, Spain. (In Spanish).

Montes-Rojas, R.T., Ospina-Noreña, J.E., Gay-Garcia, C., Rueda-Abad, C. and Navarro-Gonzalez, I. (2015). Water-Resource Management in Mexico. In Setegn, S.G. and Donoso, M.C. (Eds.) Sustainability of Integrated Water Resources Management Water Governance, *Climate and Ecohydrology* (pp. 215-244). Springer International Publishing Switzerland. ISBN 978-3-319-12194-9.

Moor, R., Van Maren, M., and Van Laarhoven, C. (2002). A controlled stable tidal inlet at Cartagena de Indias, Colombia, Terra et Aqua No. 88, Cali, Colombia. http://www.iadc-dredging.com/downloads/terra/terraet-aqua_nr88_01.pdf.

Morel, A., Schertenielb, R., and Zürbrugg, C. (2003). Alternative Environmental Sanitation Approaches in Developing Countries. EAWAG 57: 18-20.

Morgan, J.M., López, J. and Noyola, A. (1998). Matriz de Decisión para la Selección de Tecnología Relacionada con el Tratamiento de Aguas Residuales. (In Spanish). *http://www.cepis.org.pe/bvsaidis/aresidua/peru/mextar058.pdf*

Moscoso, J., Egocheaga, L., Ugaz, R. and Trellez, E. (2002). Sistemas integrados de tratamiento y uso de aguas residuales en América Latina: realidad y potencial. Centro Panamericano de Ingeniería Sanitaria y Ciencias del Ambiente (CEPIS), Panamerican Health Organization PAHO, International Development Research Centre IDRC, Canada. (In Spanish).

Mutikanga, H.E., Sharma, S.K. and Vairavamoorthy, K. (2011). Multi-criteria Decision Analysis: A Strategic Planning Tool for Water Loss Management. Water Resources Management 25, 3947-3969.

Navarro, I. and Zagmut, F. (2009). Chemical health risks. In Jimenez B. and Rose J. (Eds.). Urban Water Security: Managing Risks. *Urban Water Series* (pp 53-112) – UNESCO-IHP. ISBN 978-0-415-48567-8.

Nelson, V.I. (2008). Viewpoint: Truly Sustainable Water Infrastructure It's time to invest in next-generation decentralised technologies. *WE&T Magazine, Water Environment Federation*, September, Vol. 20, No. 9.

Nhapi, I. and Gijzen, H.J. (2005). A 3-Step Strategic Approach to sustainable wastewater management. *Water SA*, 31(1), pp. 133-140.

Novotny, V. (2008). A new paradigm of sustainable urban drainage and water management, in Oxford Roundtable Workshop on Sustainability, Paper presented at the Oxford Roundtable Workshop on Sustainability - Oxford University. pp. 1-27.

Novotny, V., and Brown P. (2007). Cities of the future: The fifth paradigm of urbanization. In Novotny, V. and Brown, P. (Eds.) Cities of the Future. Towards integrated sustainable water and landscape management. London, IWA Publishing.

Noyola, A., Morgan- Sagastume, J.M. and Guereca, L.P. (2013). Selección de Tecnología para el tratamiento de aguas residuales municipales. Universidad Autonoma de México. ISBN: 978-607-02-4822-1. (In Spanish).

OECD (2017). Diffuse Pollution, Degraded Waters: Emerging Policy Solutions, OECD Publishing, Paris. http://dx.doi.org/10.1787/9789264269064-en.

Ordoñez, J.J. (2011). Qué es una cuenca hidrológica. Cartilla Tecnica. Sociedad Geografica de Lima and Global Water Partnership GWP South America. (In Spanish).

Ostroumov, S.A. (2005). On some issues of maintaining water quality and self-purification. *Water Resource* 32(3):305–313.

Ostroumov, S.A. (2006). Biomachinery for maintaining water quality and natural water selfpurification in marine and estuarine systems: elements of a qualitative theory. *Int. Journal of Ocean and Oceanography* 1(1):111–118. ISSN 0973-2667.

Ottoson, J., and Stenström, T.A. (2003). Faecal contamination of greywater and associated microbial risks. *Water Research*, 37(3), 645-655.

Peña-Guzman, C.A., Melgarejo, J., Prats, J., Torres, A. and Martinez, S. (2017). Urban Water Cycle Simulation/Management Models: A Review. *Water*, 9, 285; doi:10.3390/w9040285.

Philip, R., Anton, A. and Loftus, A.C. (2011). Strategic Planning - Preparing for the Future. SWITCH, Training Kit Module 1. Integrated Urban Water Management in the City of the Future. Freiburg: ICLEI European Secretariat GmbH. ISBN 978-3-943107-03-6.

Posada, C. (2008). Cambio climático ¿una caja de pandora? *Hydrology,* 349, 291-301.

Public Utilities Board (2016). Innovation in Water Singapore. *Public Utilities Board*, Vol. 8 (June 2016).

Regmi, P., Lackner, S., Vlaeminck, S., Makinia, J. and Murthy, S. (2016). Nutrient Removal and Recovery: Trends and Challenges. In Global Trends & Challenges in Water Science, Research and Management. A compendium of hot topics and features from IWA Specialist Groups Second Edition. International Water Association (IWA) Alliance House, London, UK. ISBN 9781780408378, pp 91-94.

Reid, G.W. (1982). Prediction Methodology for Suitable Water and Wastewater Processes.

Rogers, P. and Hall, A.W. (2003). Effective Water Governance. TEC Background Papers No. 7, Global Water Partnership, Technical Committee, Stockholm, Sweden.

Saha, A.K. and Setegn, S.G. (2015). Ecohydrology for Sustainability of IWRM: A Tropical/Subtropical Perspective. In Setegn S.G. and Donoso M.C. (Eds.) Sustainability of Integrated Water Resources Management Water Governance, Climate and Ecology, pp. 163-178. Springer International Publishing Switzerland. ISBN 978-3-319-12194-9.

Sato, T., Qadir, M., Yamamoto, S., Endo, T. and Zahoor, A. (2013). Global, regional, and country level need for data on wastewater generation, treatment, and use. *Agricultural Water Manage*ment 130 (2013) 1– 13.

Savci, S. (2012). An Agricultural Pollutant: Chemical Fertilizer. *Environmental Science and Development*, 3(1), pp 77-79.

Seeliger, L. and Turok, I. (2013). Towards sustainable cities: Extending resilience with insights from vulnerability and transition theory. *Sustainability*, 5(5), pp. 2108-2128, DOI: 10.3390/su5052108.

Sierra, J.F. (2006). Tratamiento y reutilización de aguas grises en proyectos de vivienda de interés social a partir de humedales artificiales, Universidad de los Andes, Bogotá, Colombia. (In Spanish).

Singhirunnusorn, W. (2009). An Appropriate Wastewater Treatment System in Developing Countries: Thailand as a Case Study. PhD Thesis, Civil Engineering, University of California, Los Angeles. USA.

Sitzenfreia, R. and Raucha, W. (2014). Investigating transitions of centralized water infrastructure to decentralized solutions - an integrated approach. *Procedia Engineering* 70 (2014) 1549 – 1557.

Sobalvarro, J.A. and Batista, N.N. (1997). Propuesta para selección de procesos de tratamientos de esgotos sanitarios adecuados a ciudades de pequeño y medio porte. Universidad Federal de São Carlos. Brasil.

Tang, S.L., Wong, C.L. and Ellis, K.V. (1997). An Optimization Model for the Selection of Wastewater and Sludge Treatment Alternatives. *J. CIWEM.* 11:14-23.

Tejada-Guibert, J.A. (2015). Integrated Water Resources Management (IWRM) in a Changing World. In Setegn, S.G. and Donoso, M.C. (Eds.) Sustainability of Integrated Water Resources Management Water Governance, *Climate and Ecohydrology* (pp. 51-64). Springer International Publishing Switzerland. ISBN 978-3-319-12194-9.

Tortajada, C. (2006). Water management in Singapore. *Water Resources Development* 22(2), pp. 227-240.

Tulodziecki, D. (2011). A case study in explanatory power: John Snow's conclusions about the pathology and transmission of cholera. *Studies in History and Philosophy of Biological and Biomedical Sciences* 42(3): 306-316.

U.S. EPA (2008). Handbook for Developing Watershed Plans to Restore and Protect Our Waters. United States Environmental Protection Agency Office of Water Nonpoint Source Control Branch Washington, DC 20460 EPA 841-B-08-002.

UNEP (1998). Appropriate technology for sewage pollution control in the wider Caribbean Region. Caribbean Environment Programme, UNEP Technical Report no. 40, 1998, pp 193.

UNEP (2003). IETC Freshwater Management Series No. 7, Phytotechnologies, A Technical Approach in Environmental Management.

UNEP, GPA and UNESCO-IHE (2004). Improving Municipal Wastewater Management in Coastal Cities. A Training Manual for Practitioners. Published by Training GPA, The Hague, The Netherlands.

UNESCO - IHP (2015). Urban Waters challenges in the in the Americas. A perspective from the Academies of Sciences.

United Nations (2014). World Urbanization Prospects: The 2014 Revision, Highlights (ST/ESA/SER.A/366). United Nations, Department of Economic and Social Affairs, Population Division: New York, NY. 2.

United Nations (2015) Department of Economic and Social Affairs, Population Division. World Population Prospects: The 2015 Revision, Key Findings and Advance Tables. Working Paper No. ESA/P/WP.241.

United Nations –Water, (2015). Wastewater Management: A UN-Water Analytical Brief. UN-Water.www.unwater.org/fileadmin/user_upload/unwater_new/docs/UN-Water_Analytical_Brief_Wastewater_Management.pdf.

United Nations (2016). The Sustainable Development Goals Report, New York USA.

Vagnetti, R., Miana, P., Fabris, M. and Pavoni, B. (2003). Self-purification ability of a resurgence stream, *Chemosphere,* 52(2003), pp 1781-1795.

Vairavamoorthy, K., Eckart, J., Tsegaye, S., Ghebremichael, K. and Khatri, K. (2015). A Paradigm Shift in Urban Water Management: An Imperative to Achieve Sustainability in Setegn, S.G. and Donoso, M.C. (Eds.) Sustainability of Integrated Water Resources

Management Water Governance, *Climate and Ecohydrology* (pp. 51-64). Springer International Publishing Switzerland. ISBN 978-3-319-12194-9.

Van der Steen, P. and Howe, C. (2009). Managing Water in the City of the Future Strategic, Planning and Science. In: Reviews in *Environmental Science and Biotechnology,* 8, 115-120.

Veenstra, S., Alaerts, G.J. and Bijlsma, M. (1997). Technology Selection. Water Pollution Control - A guide to the use of water quality management principles. In Helmer, R. and Hespanhol, I. (Eds.). Published on behalf of WHO/UNEP by F & FN Spon, London, UK.

Velez, C.A., Alfonso, L., Sanchez, A., Galvis, A. and Sepulveda, G. (2014). Centinela: an early warning system for the water quality of the Cauca River. *Journal of Hydroinformatics.* 6, 1409-1424.

Von Sperling, M. (1996). Comparison Among the Most Frequently Used Systems for Wastewater Treatment in Developing Countries. *Water Science and Technology* 3(3): 59-77.

Von Sperling, M. (2005). Introdução à Qualidade das Águas e ao Tratamento de Esgotos. Departamento de Engenharia Sanitária e Ambiental (DESA), Universidad Federal de Minas Gerais (UFMG). Belo Horizonte, Brazil. Second Edition. (In Portuguese).

Von Sperling, M. (2007). Estudos e Modelagem da Qualidade da Água de Rios. Departamento de Engenharia Sanitária e Ambiental (DESA), Universidad Federal de Minas Gerais (UFMG). Belo Horizonte, Brazil. First Edition. (In Portuguese).

Wagner, I., Izydorczyk, K., Drobniewska, A., Fratczak, W. and Zalewski, M. (2007). Inclusion of Ecohydrology concept as integral component of systemic in urban water resources management: The City of Lodz Case Study, Poland. Int. Symp. New Dir. *Urban Water Manag.* 1-9.

Wagner, I. and Breil, P. (2013). The role of ecohydrology in creating more resilient cities. *Ecohydroly. Hydrobioly,* 13 (13), 113–134.

Walker, B., Hollin, C.S., Carpenter, S.R. and Kinzig, A. (2004). Resilience, adaptability and transformability in social-ecological systems. *Ecology and Society,* 9(2), 5. http://www.ecologyandsociety.org/vol9/iss2/art5/

Wang J.Q., Zhong Z. and Wu, J. (2004). Steam water quality models and its development trend, *Journal of Anhui Normal University (Natural Science),* vol. 27, no. 3, pp. 243–247.

Wang, Q, Li, S, Jia, P, Qi, C. and Ding, F. (2013). A review of surface water quality models. *The Scientific World Journal.* Article ID 231768, 7 pages. http://dx.doi.org/10.1155/2013/231768

WCED (United Nations World Commission on Environment and Development) (1987). Our Common Future.

WHO and UNICEF (2000). Global Water Supply and Sanitation Assessment 2000 Report. Report produced by World Health Organization and the United Nations Children's Fund. pp. 80. New York, USA.

WHO, UNEP and FAO (2006). Guidelines for the Safe Use of Wastewater, Excreta and Greywater. World Health Organization (WHO). France.

WHO (2016). Protecting surface water for health. Identifying, assessing and managing drinking-water quality risks in surface-water catchments.

Wilderer, P.A. (2001). Decentralized versus centralized wastewater management. In Lens, P.G. Zeeman, and G. Lettinga (Eds.). IWA Publishing.

Winpenny, J., Heinz, I., Koo-Oshima, S., Salgot, M., Collado, J., Hérnandez, F. and Torricelli, R. (2013). Reutilización del agua en agricultura: beneficios para todos, Alfabeta Artes gráficas Ed., FAO, Italia. (In Spanish).

WISE (2007) Introduction to the New Europe Union Water Framework Directive. http://ec.europa.eu/environment/water/water-framework/info/intro_en.htm.

Wittmer, I. and Burkhardt, M. (2009). Dynamics of biocide and pesticide input. In Anthropogenic micropollutants in water: impacts - risks - measures. *Federal Institute of Aquatic Science and Technology. Eawag News N° 67e*, pp. 4-11.

Wong, T.H.F and Ashley, R. (2006). International Working Group on Water Sensitive Urban Design, submission to the IWA/IAHR Joint Committee on Urban Drainage, March 2006.

Wong, T.H.F. and Brown, R. R. (2008). Transitioning to water sensitive cities: ensuring resilience through a new hydro-social contract. In: 11th International Conference on Urban Drainage. Edinburgh, Scotland, pp. 1-10.

Wong, T.H.F. and Brown, R.R. (2009). The Water Sensitive City: Principles for Practice. *Water Science and Technology*, 60(3):673-682.

WWAP (United Nations World Water Assessment Programme) (2017). The United Nations World Water Development Report 2017. Wastewater: The Untapped Resource. Paris, UNESCO.

Yang, C.T. and Kao, J.J. (1996). An Expert System for Selecting and Sequencing Wastewater Treatment Processes. *Water Science and Technology* 34(3-4): 347-353.

Zalewski, M. (2000). Ecohydrology - the scientific background to use ecosystem properties as management tools toward sustainability of water resources. *Ecol. Eng.* 16:1-8.

Zalewski, M. (2006). Ecohydrology - an interdisciplinary tool for integrated protection and management of water bodies. *Arch. Hydrobiol. Suppl.* 158/4, pp. 613-22.

Zalewski, M. and Wagner, I. (2008). Ecohydrolgy of Urban Aquatic Ecosystems for Healthy Cities. In Wagner, I, Marsalek, J. and Breil, P. (Eds.) Aquatic habitats in *sustainable Urban Water Management Science, Policy and Practice*, pp 95-106.

Zeng, G., Jiang, R., Huang, G., Xu, M. and Li, J. (2007). Optimization of wastewater treatment alternative selection by hierarchy grey relational analysis. *Environmental Management*, 82(2), 250-259.

Chapter 3
Evaluation of pollution prevention options in the municipal water cycle

Source: CVC photo file

This chapter is based on:
Galvis, A., Zambrano, D., Van der Steen, N.P. and Gijzen, H. (2014). Evaluation of pollution prevention options in the municipal water cycle. *Journal of Cleaner Production*, 66, 599-609; doi: 10.1016/j.jclepro.2013.10.057.

Abstract

The impact on water resources caused by municipal wastewater discharges has become a critical and ever-growing environmental and public health problem. In order to be able to efficiently address this problem, it is important to adopt an integrated approach that includes a decrease in and control of contamination at its source. These principles have been successfully applied in the industrial sector and now these concepts are also being applied to integrated water resources management. In this context the conceptual model of the Three Steps Strategic Approach (3-SSA) was developed, consisting of: 1) minimization and prevention, 2) treatment for reuse and 3) stimulated natural self-purification. This paper is focused on the first step. The assessment includes a case study in the expansion area of the city of Cali, Colombia (410,380 new inhabitants). The evaluation of alternatives is done using two different system boundaries: (1) reduction in water supply costs for households and the avoided costs in the infrastructure of additional sewerage and wastewater treatment facilities; and (2) only taking into account the reduction in water supply costs for households and the savings associated with the drinking water infrastructure. The alternatives of minimization and prevention were hierarchized using an analytic hierarchy process and grey relational analysis. A cost-benefit analysis was carried out to compare the highest ranked alternatives with the conventional approach, which considers a 'business as usual scenario' of high water use, end-of-pipe wastewater treatment plant and the conventional water supply system with drinking water quality for all uses. The best minimization and prevention alternatives for Cali's expansion zone were found to be those which consider double discharge toilets and the possibility of using rainwater harvesting for laundry purposes. On the other hand, the minimization and prevention alternatives considered are only viable if these are implemented in more than 20% of household units.

3.1 Introduction

To achieve sustainable urban water management, the conventional approach of high water volume and high quality for all use functions needs to be revisited. Traditionally, pollution control consists primarily of centralized and end-of-pipe solutions. Due to the high costs of this approach, it is estimated that worldwide only about 15% of all people are connected to a wastewater treatment facility that is built to provide a primary or secondary level of treatment (Bos *et al.*, 2004). The number of people connected to modern wastewater treatment facilities that include nutrient removal comprises only an estimated 2% of the world's population. It is clear that the vast majority of the indicated coverage for wastewater treatment is found in developed regions (UNEP/GPA and UNESCO-IHE, 2004). As a result, the overwhelming majority of municipal sewage is discharged untreated into rivers, lakes and coastal waters, leading to severe water quality deterioration. In fact, achieving Target 10 of the Millennium Development Goals for drinking water will lead to a further increase in sewage production, and therefore could trigger a further worsening of the already critical water quality crisis globally. A change in urban water management is necessary in order to improve the system's sustainability, and must integrate economic, social and environmental issues with practices such as integrated management of storm water, water conservation, reuse of wastewater, rational energy management, recovery of nutrients and source separation (Daigger, 2009).

Cleaner Production (CP) can be defined as the approach in which processes and activities are carried out in such a manner that the environmental impact thereof is as low as possible. As a result, the approach is now shifting from 'waste management' to 'pollution prevention and waste minimization' (Siebel and Gijzen, 2002; Veenstra *et al*, 1997). CP production concepts have been successfully applied in the industrial sector, and could help transform the urban water sector. It has been proposed that these concepts could be applied to water resources integrated management, searching for new alternatives to the limited achievements provided by end-of-pipe solutions. In this context the conceptual model of the Three Steps Strategic Approach (3-SSA) was developed, consisting of: 1) minimization and prevention, 2) treatment for reuse and 3) stimulated natural self-purification (Siebel and Gijzen, 2003; Naphi and Gijzen, 2005; Gijzen, 2006). The minimization and prevention concept refers to the reduction of residues, emissions and discharges of any production process through measures that make it possible to decrease, to economically and technically feasible levels, the amount of contaminants generated which require treatment or final disposal (Cardona, 2007; Siebel and Gijzen, 2002). The minimization proposals can be classified in three main actions (Cardona, 2007; Nhapi and Gijzen, 2005): a) reduction at source, which includes a change in consumption habits and application of low consumption devices; b) in situ recycling techniques, and c) rainwater harvesting. The first action proposes a shift to low consumption devices, such as water-saving toilets, showers and aired faucets that generate a decrease in the consumption of water, allowing for the possibility of supplying more users, without the need for new water sources and treatment capacity. The second and third actions, in situ recycling techniques, recognize new alternative water sources, such as rainwater harvesting and grey water. Lastly, the use of treated grey water is feasible for toilet flushing, the washing machine, plant watering, and the washing of floors and outdoor areas (Liu *et al.*, 2010; Mejia *et al.*, 2004; Sierra, 2006; Gijzen, 2006), golf courses, agriculture and groundwater recharge (Ottoson and Stenström, 2003).

This chapter focuses on Step 1: minimization and prevention (by applying cleaner production principles) and applies this to the case study in the city of Cali, Colombia (the expansion area). The evaluation of alternatives is done using two different system boundaries: (1) a reduction in water supply costs for households, the avoided costs in the additional drinking water infrastructure and the additional sewerage and wastewater treatment facilities; and (2) only taking into account a reduction in water supply costs for households and the savings associated with the drinking water infrastructure. The alternatives of minimization and prevention were hierarchized using an Analytic Hierarchy Process (AHP) and Grey Relational Analysis (GRA). A cost-benefit analysis was carried out to compare the highest ranked alternatives with the conventional approach, which considers a 'business as usual scenario' of high water use, end-of-pipe wastewater treatment plant and the conventional water supply system with drinking water quality for all uses.

In the holistic, integrated wastewater approach it is essential to know the impacts of particular decisions and selected strategies. An integration of technical, environmental, social, cultural, economic, policy and regulatory aspects allows for a transition from the traditional approach to one of closed and efficient processes (Zein, 2006). This approach has had gaps and has usually been focused on the wastewater treatment plant WWTP investment (end-of-pipe solutions), mainly in developed countries. Also, most of the strategies (models, guides, algorithms, among others) to support the technology selection process have been mainly oriented only towards treatment systems. Most of these tools do not consider strategic approaches such as the Three Steps Strategic Approach (3-SSA). The common selection criteria for most authors can be classified into the following factors: treatment objectives, technological aspects, costs, operation and maintenance, wastewater characteristics, demographical and socio-cultural factors, site characteristics, climate factors, environmental impact, capacity and willingness to pay, and construction aspects (Galvis *et al.*, 2006). Before selecting and investing in wastewater technology it is preferable to investigate whether pollution can be minimized or prevented (Veenstra *et al.*, 1997). Some selection models that incorporate multi-criteria analysis are: PROSAB, SANEX, WAWTTAR and PROSEL. More recent models, such as Urban Water Optioneering Tool UWOT, facilitate the selection of combinations of water-saving strategies and technologies and support the delivery of integrated, sustainable water management for new developments (Makropoulos *et al.*, 2008).

Water management is typically a multi-objective problem which makes multicriteria decision analysis MCDA a well-suited decision support tool (Hajkowicz and Collins, 2007). There is no single multi-criteria decision analysis MCDA method that can claim to be a superior method for all decision (Mutikanga *et al.*, 2011). Whilst selection of the MCDA technique is important more emphasis is need on the initial structuring of the decision problem, which involves choosing criteria and decision options (Hajkowicz and Higgins, 2008). The wastewater treatment alternative selection is a MCDA, where uncertainty, complexity and hierarchy need to be considered. (Zeng *et al.*, 2007) propose a multi-criteria analysis methodology including: AHP and (GRA). AHP is useful for handling multiple criteria and objectives in the decision-making process. The GRA is a measurement method in grey system theory that analyzes uncertain relations between one main factor and all the other factors in a given system (Liu *et al.*, 2005; Tosun and Pihtili, 2010). The hierarchy GRA combines the traditional GRA with the idea of the hierarchy of the AHP. It enables a more effective evaluation than just the mono level-based evaluation. The different levels of importance of the criteria are reflected through weighting factors to avoid subjectivity and randomness. In addition, the quantified evaluating scale, namely the integrated grey relational grade, makes the wastewater treatment alternative selection more comparable and comprehensive. Grey system theory was developed by Deng (1982) and has been successfully applied in engineering prediction and control, social and economic system management, and environmental system decision making in recent years.

This chapter aims to identify and validate ways to maximize the benefits of the strategy (3-SSA) in the municipal water cycle and to provide the tools and approach for the selection of viable and effective alternatives under Step 1. The research presents the potential usage of AHP + GRA in the hierarchies of water-saving alternatives in households, leading to domestic wastewater pollution minimization and prevention. This selection methodology includes a cost-benefit analysis (CBA) among the highest-rated alternatives (AHP + GRA results) and a comparison with the conventional approach, which considers a 'business as usual scenario' of high water use, end-of-pipe wastewater treatment plant and the conventional water supply system with drinking water quality for all uses.

The Three-Step Approach as compared to more conventional approaches may lead to a more cost-effective policy choice, assuming similar health gains (Bos *et al.*, 2004). According to WHO (2004), investing in sanitation and water supply projects provides economic benefits due to the fact that for each US$ invested, there is an economic benefit ranging between US$ 3 and US$ 34, depending on the region. These economic benefits include impacts on: population health, environment, agriculture, industry, economy, tourism, etc. (OPS, 2008). This study uses the incremental cost-benefit analysis and it does not consider the common costs and benefits to compare the approaches. It also did not consider benefits of minimization and prevention in relation to the other two steps of 3-SSA: renewable energy production; water recovery and reuse potential; nutrient recovery and reuse, including savings on fertilizer and environmental benefits.

The paper describes the study area and presents the methodology for identification and characterization of minimization and prevention alternatives. Then, the multicriteria analysis is described AHP + GRA and the basic criteria to CBA is indicated. The results describe 10 minimization and prevention alternatives. These alternatives are ranked as a result of applying AHP + GRA. Then, the 4 best alternatives are compared with conventional approach using the

CBA. A sensitivity analysis is presented considering different combinations of percentages of single-family and multifamily dwellings and different percentages of households implementing prevention and minimization alternatives.

3.2 Methods

3.2.1 Study area: expansion area Cali-Jamundi corridor

The study was carried out in the expansion area of the city of Cali (Figure 3.1). It is a future development (still do not exist) with an area of 1,669 ha and is located in the Jamundí and Lili River basins, between 955 and 1,030 meters above sea level. Slopes may vary between 3% and 15%, facilitating water drainage into the Cauca, Jamundí and Lilí rivers. Cali has an average temperature range between 23°C and 25°C, with bimodal behaviour in terms of precipitation and evaporation. Monthly average precipitation over multiple years in the expansion area is 122 mm. Maximum rainfall is 196 mm in April and the lowest rainfall is 51 mm in July.

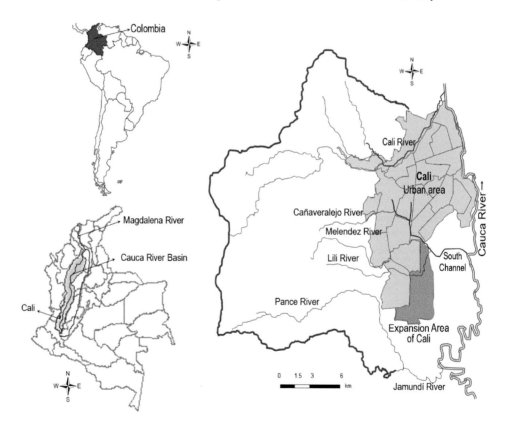

Figure 3.1 Cali, Colombia and its expansion area

The expansion area is mainly characterized by residential land use (1,358 ha), with a gross density of 302 inhabitants per ha, and a total of 410,380 inhabitants (EMCALI and Hidroccidente S.A., 2006). 15% of households are single-family houses and 85% multi-family apartments. There is an average of 4 persons per household in strata 3 to 6 of the city (Socioeconomic strata in Colombia are: 1, 2, 3, 4, 5 and 6. The lowest corresponds to stratum 1) (Departamento Administrativo de Planeacion Municipal, 2008), and a total of 102,595 households (15,784 single-family and 86,811 multi-family type). The solution proposed by a local consulting company is based on end-of-pipe solutions. This proposal is used in this study and it has been labelled the 'conventional approach'. The following data is used: average supply of 225 L/inhabitant/day and drinking water losses of 25%. In 2030, the water demand will be 1,067 L/s. Water supply will be obtained through the matrix network expansion towards the southern part of Cali, and therefore will require pumping. The sewer system is separate and the wastewater production is 170 L/inhabitant/day. Two secondary wastewater treatment plants (WWTPs) are planned (in year 1 and year 10). The technology used is a high rate anaerobic lagoon and a facultative lagoon.

3.2.2 Identification and characterization of alternatives

The identification included a literature review, consultation with experts and local and external market research for devices that may contribute to a decrease in water consumption. Information about local characteristics was used, in terms of environmental and household conditions (single and multifamily households), and initial investment and operational costs. A group of alternatives was considered in the preliminary selection, and these were compared according to their initial investment and operation and maintenance costs. Based on such a selection, a social consultation on low consumption devices was made through 167 surveys carried out with individuals interested in buying houses in the study area. The survey assessed: type of preferred toilet and levels of knowledge and acceptability of the use of grey water and rainwater. The different alternatives were analysed according to water demand, wastewater production, rainwater excess and a small reduction in BOD and TSS loads generated. The evaluation of these alternatives is done using two different system boundaries: (1): reduction in water supply costs for households and the avoided costs in the infrastructure of additional sewerage and wastewater treatment facilities; and (2) only taking into account the reduction in water supply costs for households and the savings associated with the drinking water infrastructure.

Characterization of alternatives included the layouts, pre-dimensioning and costs of the main water supply and sewage networks and WWTPs, as well as drinking water pumping requirements. The drinking water flow demand and wastewater production were also included (Zambrano, 2012). Initially, the assumption used in the study was: 70% (71,817) of households would apply systems that included minimization and prevention alternatives.

3.2.3 Analytic Hierarchy Process (AHP)

The main advantage of the AHP is its ability to rank choices in the order of their effectiveness in meeting conflicting objectives. The main limitation of AHP is that it is integrated with a comprehensive axiomatic scheme that is a heuristic method which can obtain reasonable results to multi-criteria complex decisional problems (Romero, 1997). In this study the application of this model included four analysis levels (criteria): environmental, economic, social and technical. The relevance of the criteria was identified through consultation of local stakeholders. The surveys provided information on the relevance of each criteria compared to others in order to obtain the pair comparison matrix (Saaty, 1990; Romero, 1997; Zeng *et al.*, 2007). The consistency ratio (CR) was verified by calculating the consistency index (CI) and random average index (RI) (Sanchez, 2003; Saaty, 2008). The same weighting was used for the indicators identified within each criterion.

Investment costs of the housing infrastructure and external infrastructure, for each alternative, were estimated based on the drinking and wastewater flows, and the treatment plant capacity. In order to identify the complexity level, local professional experts in the water sector were consulted. The institutional support indicator was set through 19 surveys filled out by active members of water sector institutions. For social acceptance some people (167) interested in purchasing (potential buyers) a household in the study area with similar characteristics to the study area were surveyed. A casual sampling type (Pimienta, 2000) was used. A selection matrix was made with the values obtained for each indicator. All data were normalized to a scale of 0 to 1 using the following expressions (Ye, 2003, cited by Zeng *et al.*, 2007).

For indicators to minimize the normalized data can be obtained by Equation (3.1), while for indices to maximize, the normalized data can be obtained by Equation (3.2):

$$Xij = \frac{min_i\{s_i(j)\}}{s_i(j)} \quad (3.1) \quad Xij = \frac{s_i(j)}{max_i\{s_i(j)\}} \quad (3.2)$$

i:	Alternative;
j:	Criterion
Xj:	Normalized value of alternative i under criterion j
Si (j):	Value of alternative i under criterion j
min Si(j):	Minimum value of benefit in the criterion j in all alternatives
max Si(j):	Maximum value of benefit in the criterion j in all alternatives

3.2.4 The hierarchy GRA procedure

The hierarchy GRA can be implemented in three steps (Zeng *et al.*, 2007). *Step* 1: is to calculate the primary grey relational coefficient matrix (n x m) $\xi_{0i}(j)$ (Equation 3.3), while each resultant element in the matrix represents the relational coefficients between the reference alternative and a given optional alternative for a given criteria.

$$\varepsilon_{0i}(j) = \frac{0.5 \max_i \{\max_j |x_{oj} - x_{ij}|\}}{\left|x_{oj} - x_{ij} + 0.5 \max_i \{\max_j |x_{oj} - x_{ij}|\}\right|} \qquad (3.3)$$

i: Technological alternatives, $i = 1,2, \dots, m$.
j: Criteria, $j = 1,2, \dots, n$.

The primary grey relational coefficient matrix for all the indices of the optional alternatives can be denoted with Equation 3.4.

$$G = \begin{vmatrix} g_{c_1} \\ g_{c_2} \\ \dots \\ g_{c_n} \end{vmatrix} = \begin{vmatrix} \xi_{01}(1) & \xi_{02}(1) & \dots & \xi_{0m}(1) \\ \xi_{01}(2) & \xi_{02}(2) & \dots & \xi_{0m}(2) \\ \dots & \dots & \dots & \dots \\ \xi_{01}(n) & \xi_{02}(n) & \dots & \xi_{0m}(n) \end{vmatrix} \qquad (3.4)$$

g_{ck} $(k=1, 2, \dots, s)$ represent the grey relational coefficient vector for to the indices subject to k_{th} criterion C_K (Equation 3.5):

$$g_{C_k} = \begin{vmatrix} \xi_{01}(p) & \xi_{02}(p) & \dots & \xi_{0m}(p) \\ \xi_{01}(p+1) & \xi_{02}(p+1) & \dots & \xi_{0m}(p+1) \\ \dots & \dots & \dots & \dots \\ \xi_{01}(q) & \xi_{02}(q) & \dots & \xi_{0m}(q) \end{vmatrix} \qquad (3.5)$$

$I_p, I_{p+1}, \dots, I_q (1 \le p \le q \le n)$ Are the indices to the k_{th} *criterion* C_k. Hence, elements in each column $\xi_{0i}(p)$, $\xi_{0i}(p+1)$, $\dots$, $\xi_{0i}(q)$ representing the relational coefficient between reference alternative S_0 and optional alternative S_i.

Step 2 is to calculate the secondary grey relational coefficient matrix $(s \times m)$ while each row vector of the matrix represents the relational coefficients between the reference alternative and the given option for specific criterion. According the weighed primary relational coefficient vector of the indices subject to criterion C_k can be obtained (Equation 3.6).

$$\delta_{C_k} = W_{C_k} g_{C_k} = \left(W_{I_p}, W_{I_{p+1}}, \dots, W_{I_q}\right) \times \begin{vmatrix} \xi_{01}(p) & \xi_{02}(p) & \dots & \xi_{0m}(p) \\ \xi_{01}(p+1) & \xi_{02}(p+1) & \dots & \xi_{0m}(p+1) \\ \dots & \dots & \dots & \dots \\ \xi_{01}(q) & \xi_{02}(q) & \dots & \xi_{0m}(q) \end{vmatrix} \qquad (3.6)$$

$$= \left(\delta_{C_k}(1), \delta_{C_k}(2), \dots, \delta_{C_k}(m)\right)$$

Similarly, Equation (3.7) is used in order to get the corresponding weighed primary grey relational coefficient vector for any other criteria on the criterion level.

$$\mathbf{G}_{weighed} = \begin{vmatrix} \delta_{c_1}(1) & \delta_{c_1}(2) & ... & \delta_{c_1}(m) \\ \delta_{c_2}(1) & \delta_{c_1}(2) & ... & \delta_{c_1}(m) \\ ... & ... & ... & ... \\ \delta_{c_s}(1) & \delta_{c_s}(2) & ... & \delta_{c_s}(m) \end{vmatrix} \qquad (3.7)$$

With data normalization of $\mathbf{G}_{weighed}$, it is possible to improve the data comparability (Equations 3.1 and 3.2). By scaling the resulting $\delta_{c_k}(i)$ $(i = 1,2,...,m; k = 1,2,...,s)$, the normalized weighed primary grey relational coefficient matrix is then obtained (Equation 3.8)

$$\mathbf{G}'_{weighed} = \begin{vmatrix} \delta'_{c_1}(1) & \delta'_{c_1}(2) & ... & \delta'_{c_1}(m) \\ \delta'_{c_2}(1) & \delta'_{c_2}(2) & ... & \delta'_{c_2}(m) \\ ... & ... & ... & ... \\ \delta'_{c_s}(1) & \delta'_{c_s}(2) & ... & \delta'_{c_s}(m) \end{vmatrix} \qquad (3.8)$$

Where $\delta'_{c_k}(i)$ $(i = 1,2,...,m; k = 1,2,...,s)$ is the grey relational coefficient resulting from the normalization of $\delta_{c_k}(i)$.

Again Equation (3.3) is used to obtain the grey relational coefficients between the reference and optional alternatives for a certain criterion. Thus, secondary grey relational grade vector can be obtained as follow:

$$\mathbf{G}_C = \begin{vmatrix} \xi_{c_1}(1) & \xi_{c_1}(2) & ... & \zeta_{c_1}(m) \\ \xi_{c_2}(1) & \xi_{c_1}(2) & ... & \zeta_{c_1}(m) \\ ... & ... & ... & ... \\ \xi_{c_s}(1) & \xi_{c_s}(2) & ... & \zeta_{c_s}(m) \end{vmatrix} \qquad (3.9)$$

Step 3. In the last step the integrated relational grade row vector (1 x *m*) is calculated (Equation 3.10).

$$\varepsilon = \mathbf{W}_C \times \mathbf{G}_C = (W_{C_1}, W_{C_2}, ..., W_{C_k}, ..., W_{C_s}) \times \begin{vmatrix} \xi_{c_1}(1) & \xi_{c_1}(2) & ... & \zeta_{c_1}(m) \\ \xi_{c_2}(1) & \xi_{c_1}(2) & ... & \zeta_{c_1}(m) \\ ... & ... & ... & ... \\ \xi_{c_s}(1) & \xi_{c_s}(2) & ... & \zeta_{c_s}(m) \end{vmatrix} = (\varepsilon_1, \varepsilon_2, ..., \varepsilon_m) \quad (3.10)$$

3.2.5 Cost Benefit Analysis (CBA)

The best alternatives were ranked by CBA (Miranda, 2000; Brent, 2006) and compared with the conventional approach, which considers 6-liter toilets and drinking water used for all uses. This comparison was made for scenarios 1 and 2, which differ in their system boundaries. Environmental and economic benefits were calculated. Common benefits were not taken into consideration. A constant demographic growth rate for a period of twenty (20) years, and a project horizon for the cost-benefit evaluation of 30 years were adopted, as well as a social discount rate of 11% (Ministerio de Ambiente, Vivienda y Desarrollo Territorial, 2010).

3.3 Results

3.3.1 Identification and preliminary selection of minimization and prevention alternatives

The identification and characterization (Table 3.1) included: consultation with experts, literature review (Velasquez, 2009; Ghisi and Mengotti de Oliveira, 2007), information from local and external market (Corona Organization; Global Business Alliance GBA) of low consumption devices (El Espectador, 2009). Information from existing residential units located close to study area was also used. The development is planned in the expansion area of Cali. Distribution networks (drinking and grey water uses) are separated.

Table 3.1 Minimization and prevention alternatives for the expansion area of Cali

Type of toilet	Alternative	Single-family households [a]	Multi-family households [b]
WC dual flush	A	• Drinking water for all uses	• Drinking water for all uses
	B	• Grey water for toilet flushing & garden irrigation • Drinking water for laundry	• Grey water for toilet flushing & garden irrigation
	C	• Grey water for toilet flushing & garden irrigation • Rainwater harvesting for laundry	• Grey water & rainwater harvesting for cleaning communal areas
	D	• Grey water for toilet flushing & garden irrigation • Rainwater harvesting for laundry	• Drinking water for toilet flushing
	E	• Drinking water for toilet flushing • Drinking water-rainwater harvesting for laundry • Rainwater harvesting for garden irrigation	• Rainwater harvesting for garden irrigation and cleaning communal areas
WC 2.3 L	F	• Drinking water for all uses	• Drinking water for all uses
	G	• Grey water for toilet flushing & garden irrigation • Drinking water for laundry	• Grey water for toilet flushing & garden irrigation • Grey water & rainwater harvesting for cleaning communal areas
	H	• Grey water for toilet flushing & garden irrigation • Rainwater harvesting for laundry	
	I	• Drinking water for toilet flushing and laundry • Rainwater harvesting for garden irrigation	• Drinking water for toilet flushing
	J	• Drinking water for toilet flushing • Drinking water-rainwater harvesting for laundry • Rainwater harvesting for garden irrigation	• Rainwater harvesting for garden irrigation and cleaning communal areas

[a] All alternatives for single-family households use drinking water for kitchen, sink and shower.
[b] All alternatives for multi-family households use drinking water for kitchen, sink, shower and laundry.

Water demand calculations per household (considering an average of 4 people/household) assumed the following (m³/household/month): kitchen (1.2 m³); sinks (1.3 m³); showers (5.4 m³); laundry and housekeeping (2.5 m³); toilets (2.7 m³ for dual flush; 1.38 m³ for high efficiency (2.3 liter toilet); garden irrigation and others (2.0 m³ in single-family households and 0.6 m³ in multi-family households). The Figure 3.2 shows the wastewater and drinking water flows corresponding to different minimization and prevention alternatives.

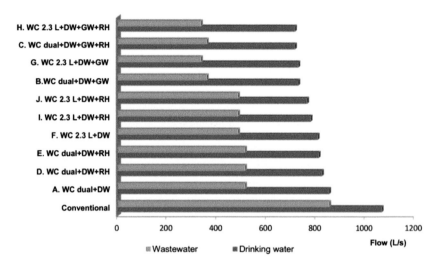

WC dual: dual flush toilet; **WC 2.3 L**: high efficiency toilet; **DW**: drinking water; **GW**: grey water; **RH**: rainwater harvesting; **Conventional approach**: toilet 6 L per flush; drinking water for all uses

Figure 3.2 Total drinking water and wastewater flows from the study area for different alternatives

3.3.2 Analytic Hierarchy Process (AHP) and Grey Relational Analysis (GRA)

Economic criteria
For the *external infrastructure,* costs are especially associated with main water supply networks and pumping stations; works associated with storm water management, sewerage systems and WWTPs. These costs are calculated based on the drinking water flow requirements and wastewater produced. Capital costs and an operation and maintenance (O&M) index are obtained considering that these costs are proportional to flow, while the range of variation of the flow between the different alternatives is small. For the *internal infrastructure,* the additional costs generated by the implementation of low consumption devices were included.

For toilets in the conventional approach a 6 L low-consumption toilet with a cost of €64 was considered. The 2.3 L toilet costs €160, while a WC dual flush costs €86. Investment costs associated with the use of rainwater harvesting and grey water correspond to €86.4 and €161 for multi-family households, and €355 and €395, respectively, for single-family households. In terms of operation and maintenance requirements of the technology, a cost of €40/year for each system (grey water and rainwater harvesting) is calculated. These costs are for periodic maintenance every four months. Other costs included are expenses caused by pump replacements every three years throughout the life cycle of the project.

Environmental criteria
The reduction of drinking water consumption is estimated between 20 and 33%, based on demand proposed for conventional water supply for the study area, using a flow of 1,067 L/s (EMCALI and Hidroccidente, 2006). A small BOD and TSS removal percentage were taken into account. The values vary between 0 and 8% for BOD and 0 and 9% for TSS. As far as the removal of BOD and TSS is concerned, annual contributions of domestic wastewater and excess storm water collected from house roofs were analysed, taking into account the fact that there are seven months of rain. Load reduction (Table 3.2) is calculated based on the flow and loads considered in the conventional approach (854 L/s; 204 mg/L of BOD and 336 mg/L of TSS). For runoff, 24 mg/L of BOD and 261 mg/L of TSS (Wanielista, 1993 cited by Navarro, 2007) were considered.

Table 3.2 Scores of indicators of minimization and prevention alternatives - Scenario 1

Criteria	Index	Alternative									
		A	B	C	D	E	F	G	H	I	J
Economic	Cost[a] external infrastructure	1.8	1.4	1.3	1.6	1.5	1.8	1.4	1.3	1.5	1.4
	Cost[b] housing infrastructure	11.3	80.6	94.5	44.0	44.0	49.26	11.9	132.5	82.0	82.0
Environm.	Decrease water demand	0.20	0.32	0.33	0.23	0.24	0.24	0.32	0.33	0.27	0.28
	Removal efficiency BOD	0%	8%	8%	0%	0%	0%	5%	5%	0%	0%
	Removal efficiency TSS	0%	8%	9%	1%	2%	0%	5%	5%	1%	2%
Technical	Level of complexity	0.6	1.6	1.7	0.8	1.1	0.6	1.6	1.7	0.8	1.1
Social	Institutional support	0.75	0.28	0.23	0.62	0.36	0.25	0.09	0.08	0.21	0.12
	Social acceptance	0.53	0.52	0.47	0.53	0.47	0.41	0.41	0.36	0.41	0.35

[a]: Capital and O&M cost. Relative values; [b] Values in thousands of Euros
(DW) Drinking Water; (GW) Grey Water;   (RW) Rainwater harvesting
A. WC dual+DW;   B. WC dual+DW+GW+RH;   C. WC dual+DW+GW+RH;
D. WC dual+DW+RH;   E. WC dual+DW+RH   F. WC 2.3L+DW;
G. WC 2.3 L+DW+GW+RH   H. WC 2,3L+DW+GW+RH;   I. WC 2.3L+DW+RH;
J. WC 2.3L+DW+RH

Technical and social criteria
The level of complexity for each combination of uses and sources was identified and a value ranking between 0 and 1 was assigned. Institutional support was identified through surveys of individuals active in water management and urban planning in Cali. Social acceptance was identified through surveys of potential house buyers in the study zone. The results are shown in the Table 3.2. $\lambda_{max} = 4.167$, The consistency CI is 0.0557 and the random average consistency index (RI) is 0.9 when the number of criteria is 4 (Saaty, 2008). Therefore, the consistency ratio (CR) is 0.062, less than 0.1, and the results pass the test of consistency. Figure 3.3 shows the hierarchy system for the selection of minimization and prevention alternatives.

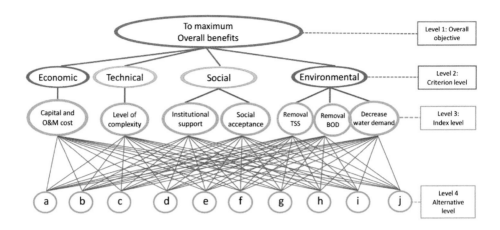

Figure 3.3 Hierarchy system for the selection of minimization and prevention alternatives

Frequency comparison of pairs and weightings of each criterion and the normalization are show in the Table 3.3.

Table 3.3 Frequency comparison of pairs and weightings of each criterion and the normalization

Criteria	Environmental	Economic	Technical	Social	W	W Unitary
Environmental	1	3	1	1	⌈1.38⌉	⌈0.32⌉
Economic	1/3	1	1	1	0.78	0.19
Technical	1	1	1	1	1.00	0.24
Social	1	1	1	1	⌊1.00⌋	⌊0.25⌋
					Σ=4.167	1.0

The Table 3.4 to Table 3.9 present the calculations for the AHP-GRA selection process. According to these calculations the best option is the alternative C, which includes dual flush toilets. The group best ranked includes the alternatives B, C, G, H, which have in common gray water use. In this group there are two blocks: C, B (dual flush toilets) and G, H (WC 2.3 L). Options C and H have in common the use of rainwater in single houses.

Table 3.4 Normalized data of each alternative - Scenario 1

Criteria	Index	A	B	C	D	E	F	G	H	I	J
Economic	Capital and O&M cost external infrastructure	0.62	0.83	0.96	0.72	0.80	0.64	0.85	1.00	0.76	0.86
	Capital and O&M cost in housing inf.	1.00	0.37	0.36	0.40	0.40	0.39	0.36	0.35	0.37	0.37
Environ.	Decrease water demand	0.56	0.93	1.00	0.62	0.65	0.66	0.93	1.00	0.74	0.78
	Removal efficiency load BOD	0.33	0.98	1.00	0.34	0.34	0.33	0.53	0.54	0.34	0.34
	Removal efficiency load TSS	0.33	0.89	1.00	0.37	0.38	0.33	0.53	0.57	0.37	0.38
Technical	Level of complexity	1.00	0.44	0.44	0.67	0.52	1.00	0.44	0.44	0.67	0.52
Social	Institutional support	1.00	0.44	0.42	0.75	0.49	0.43	0.36	0.36	0.41	0.37
	Social acceptance	1.00	0.97	0.80	0.99	0.81	0.69	0.68	0.61	0.69	0.60

Table 3.5 Primary grey relational coefficients of alternatives for sub-criteria - Scenario 1

Criteria	Index	Alternative									
		A	B	C	D	E	F	G	H	I	J
Economic	Capital and O&M cost ext. infrastructure	0.62	0.83	0.96	0.72	0.80	0.64	0.85	1.00	0.76	0.86
	Capital and O&M cost in housing inf.	1.00	0.37	0.36	0.40	0.40	0.39	0.36	0.35	0.37	0.37
	Decrease water demand	0.56	0.93	1.00	0.62	0.65	0.66	0.93	1.00	0.74	0.78
Environ.	Removal efficiency BOD	0.33	0.98	1.00	0.34	0.34	0.33	0.53	0.54	0.34	0.34
	Removal efficiency TSS	0.33	0.89	1.00	0.37	0.38	0.33	0.53	0.57	0.37	0.38
Technical	Level of complexity	1.00	0.44	0.44	0.67	0.52	1.00	0.44	0.44	0.67	0.52
Social	Institutional support	1.00	0.44	0.42	0.75	0.49	0.43	0.36	0.36	0.41	0.37
	Social acceptance	1.00	1.00	1.00	1.00	1.00	1.00	1.00	1.00	1.00	1.00

Table 3.6 Resultant primary grey relational coefficients for criterion level – Scenario 1

Criteria	A	B	C	D	E	F	G	H	I	J
Economic	0.81	0.60	0.66	0.56	0.60	0.52	0.60	0.68	0.57	0.61
Environmental	0.41	0.93	1.00	0.44	0.46	0.44	0.66	0.70	0.48	0.50
Technical	1.00	0.44	0.44	0.67	0.52	1.00	0.44	0.44	0.67	0.52
Social	1.00	0.70	0.61	0.87	0.65	0.56	0.52	0.48	0.55	0.49

Table 3.7 Normalized weighted primary grey relational coefficients - Scenario 1

Criteria	A	B	C	D	E	F	G	H	I	J
Economic	0.64	0.87	0.78	0.92	0.87	1.00	0.86	0.77	0.92	0.85
Environmental	0.41	0.93	1.00	0.44	0.46	0.44	0.66	0.70	0.48	0.50
Technical	0.44	0.98	1.00	0.65	0.83	0.44	0.98	1.00	0.65	0.83
Social	1.00	0.70	0.61	0.87	0.65	0.56	0.52	0.48	0.55	0.49

Table 3.8 Secondary grey relational coefficients - Scenario 1

Criteria	A	B	C	D	E	F	G	H	I	J
Economic	0.451	0.687	0.576	0.797	0.688	1.000	0.676	0.559	0.781	0.656
Environmental	0.333	0.812	1.000	0.346	0.353	0.346	0.468	0.499	0.363	0.373
Technical	0.344	0.939	1.000	0.461	0.638	0.344	0.939	1.000	0.461	0.638
Social	1.000	0.499	0.431	0.692	0.457	0.401	0.381	0.364	0.395	0.365

Table 3.9 The integrated grey relational grade for each alternative – scenarios 1 and 2

Criteria	Scenario	A	B	C	D	E	F	G	H	I	J
Economic	1	0.08	0.13	0.11	0.15	0.13	0.19	0.13	0.10	0.15	0.12
	2	0.08	0.14	0.11	0.13	0.11	0.19	0.14	0.11	0.13	0.10
Environmental	1	0.11	0.26	0.32	0.11	0.11	0.11	0.15	0.16	0.12	0.12
	2	0.11	0.26	0.32	0.11	0.11	0.11	0.15	0.16	0.12	0.12
Technical	1	0.08	0.23	0.25	0.11	0.16	0.08	0.23	0.25	0.11	0.16
	2	0.08	0.23	0.25	0.11	0.16	0.08	0.23	0.25	0.11	0.16
Social	1	0.25	0.12	0.11	0.17	0.11	0.10	0.09	0.09	0.10	0.09
	2	0.25	0.12	0.11	0.17	0.11	0.10	0.09	0.09	0.10	0.09
Σ	1	**0.52**	**0.74**	**0.78**	**0.54**	**0.51**	**0.48**	**0.60**	**0.60**	**0.47**	**0.49**
	2	**0.52**	**0.75**	**0.78**	**0.53**	**0.49**	**0.48**	**0.62**	**0.61**	**0.46**	**0.47**

(DW) Drinking Water; (GW) Grey Water;   (RW) Rainwater harvesting
A. WC dual+DW;   B. WC dual+DW+GW+RH;   C. WC dual+DW+GW+RH;
D. WC dual+DW+RH;   E. WC dual+DW+RH   F. WC 2.3L+DW;
G. WC 2.3 L+DW+GW+RH   H. WC 2,3L+DW+GW+RH;   I. WC 2.3L+DW+RH;   J. WC 2.3L+DW+RH

3.3.3 Minimization and prevention versus the conventional approach

Avoided costs in water supply and sanitation systems within houses due to implementation of minimization and prevention alternatives (B1). This includes cost reductions related to volumes used, purification and distribution costs (pumping, energy, chemical products and replacements); wastewater collection and treatment costs and payment of tax for water use and tax for wastewater discharge into the sewer system (Table 3.10)

Table 3.10 Avoided costs in water supply and sanitation systems within houses (B1) (values in €)

Alternative	NPV	Year						
		1	2	5	10	15	20	30
B	48,855,011	0	758,011	3,032,045	6,822,102	10,612,158	14,402,214	15,160,226
C	50,702,564	0	786,677	3,146,708	7,080,093	11,013,479	14,946,864	15,733,541
G	48,855,011	0	758,011	3,032,045	6,822,102	10,612,158	14,402,214	15,160,226
H	50,702,564	0	786,677	3,146,708	7,080,093	11,013,479	14,946,864	15,733,541
Conv. approach	0	0	0	0	0	0	0	0

Avoided costs for the external infrastructure investment due to implementation of minimization and prevention alternatives (B2) were: water supply network, pumping stations, sewage system and WWTP. Investments for the main water supply and sewage system networks are allocated to the first year. The drinking water pump station is planned to be constructed in steps, every 5 years. Two WWTPs are planned: one to be constructed in Year 1 and the second one in Year 10. The Table 3.11 shows a summary of the avoided costs corresponding to (B2) for the scenarios (1) and (2). The values in the table apply to alternatives B, C, G and H, because total drinking water and wastewater flows for each of these alternatives are similar.

Table 3.11 Avoided costs due to implementation of min and prevention M&P (B2) (values in €)

Item	Scenario	NPV	Year 1	Year 5	Year 10	Year 15	Year 20
M&P alternatives[a]							
Drinking water distribution	1	2,651,448	2,943,107	-	-	-	-
network	2	2,651,448	2,943,107	-	-	-	-
Drinking water pumping station	1	113,622	31,111	31,111	115,556	28,889	164,444
	2	113,622	31,111	31,111	115,556	28,889	164,444
Sanitary sewer system	1	806,427	895,134				
Wastewater treatment plant	1	1,371,814	1,094,749	-	1,094,749	-	-
Cash flow M&P alternatives	1	4,943,311	4,964,101	31,111	1,210,305	28,889	164,444
	2	2,765,070	2,974,218	31,111	115,556	28,889	164,444
Conventional approach							
Drinking water distribution	1	2,862,652	3,177,544	-	-	-	-
network	2	2,862,652	3,177,544	-	-	-	-
Drinking water pumping station	1	157,426	46,667	115,556	28,889	28,889	246,667
	2	157,426	46,667	115,556	28,889	28,889	246,667
Sanitary sewer system	1	955,980	1,061,138	-	-	-	-
Wastewater treatment plant	1	4,637,973	2,708,268	-	2,708,268	-	-
Cash flow conventional approach	1	8,614,031	6,993,617	115,556	2,737,157	28,889	246,667
	2	3,020,078	3,224,211	15,556	28,889	28,889	246,667
Total savings	1	3,670,720	2,029,516	84,444	1,526,853	-	82,222
	2	255,008	249,993	84,444	-86,667	0	82,222

[a] Apply to each of the alternatives B, C, G and H

Cost for the implementation of minimization and preservation alternatives (Table 3.12) are associated with the additional internal network infrastructure, including initial investment and operation and maintenance costs. For Alternative B, initial investments are: €439 for single households (low-consumption power device: €44; grey water system: €395) and €291 for the multiple households (low-consumption device: €44; rain water harvesting: €86; grey water system: €161).

Table 3.12 Costs of implementing min. and prevention alternative B (Thousands of €)

Item	NPV	1	2	5	10	15	20	30
Initial Inv. internal network. of water & san.	11,913	1,496	1,496	1,496	1,496	1,496	1,496	0
Operation & maintenance	18,515	0	287	1,149	2,585	4,021	5,458	5,745
Replacement	1,301	0	0	68	205	274	411	411
Total cost	31,729	1,496	1,783	2,713	4,286	5,791	7,365	6,156

3.3.4 Economic feasibility indicators

The main differences in cash flow and scenarios 1 and 2 (Figure 3.4) correspond to investment in WWTPs (years 1 and 10). In Year 1, the benefit is associated with savings made in the initial investment of external network and water supply and sanitation infrastructure because of the smaller dimensions of pipelines or equipment. There is a noticeable decrease in the amount of work required, substantially lowering costs. After Year 1, cash flow in both cases is negative due to the incorporation of new users and investment requirements for internal networks. This situation continues until around Year 8, when the number of users, implementing the minimization and prevention alternatives, receives the benefit of water supply and sewage service savings, reaching a break-even point with costs for newly connected users. In Year 10, a benefit is obtained due to savings in the second water treatment plant. Likewise, tariff benefits increase until Year 21, when the study area is totally populated. The cost-benefit ratio for alternatives B, C, G and H (scenarios 1 and 2) is presented in the Table 3.13

Table 3.13 Economic indicators of different minimization and prevention alternatives

Criteria	Scenario	Alternative B	Alternative C	Alternative G	Alternative H
$NPV_{Benefit} / NPV_{cost}$	1	1.14	1.22	1.09	1.08
	2	1.07	1.15	1.03	1.02

3.3.5 Sensitivity analysis

420 Combinations were analysed, as per 7 different percentages of the household types (single and multi-family), 6 different percentages of households implementing minimization and prevention strategies, and 10 different minimization and prevention alternatives (A, B, C, D, E, F, G, H, I, J). The break-even point of the feasibility of the minimization and prevention strategies is approximately 20% for scenarios (1) and (2). For these two scenarios, in urban models with a greater number of single family households, B is the most feasible alternative,

while if the model has a larger number of multi-family households, the best alternative is C. The urban models generating the greatest benefits are those corresponding to 100% single-family homes (alternatives B, G, H) and to 100% multi-family homes (alternative C). When reviewing the eligibility order for different alternatives, for scenarios 1 and 2, it was found that alternatives B, C, G and H are the ones most frequently ranked in the first four positions.

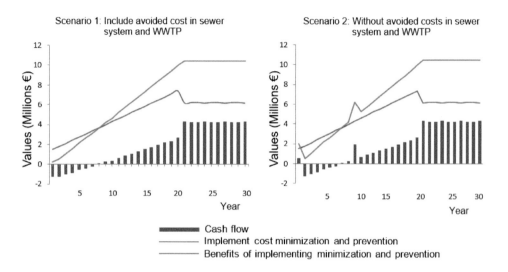

Figure 3.4 Net benefit of Alternative B versus the conventional approach, scenarios (1) and (2)

3.4 Discussion

The strength of the Three-Step Strategic Approach (3-SSA) (Nhapi and Gijzen, 2005; Gijzen, 2006) is based on ensuring the implementation of all its steps. The benefits arising from interventions in one step (e.g. under Step 1) will lead to additional savings in the subsequent steps; e.g. substantial reduction in drinking water use (step 1) yields more concentrated wastewater, which improves efficiency in treatment systems and provides more opportunities for resource recovery in Step 2. The identification and validation of the advantages of 3-SSA can be studied for each step, considering the possible borders between each. For the first step, it is necessary to choose the best minimization and prevention alternative, considering the technical, social, environmental and economic issues. This selection is a multiple-objective decision-making process.

The GRA has been used in this study to compare and to evaluate the minimization and prevention alternatives. The hierarchy GRA combines the traditional GRA with the idea of the hierarchy of the AHP (Zeng *et al.*, 2007). AHP+GRA analysis can be improved by considering the uncertainty using fuzzy logic instead of the Boolean logic (Kahraman, 2008). In this methodology a 'grey number' belongs to a range (lower and upper bound) instead of crisp value. An economic evaluation of the costs and benefits of minimization and prevention versus the conventional approach will determine the viability of this first step, even without considering the impacts of the other two steps. CBA complements the AHP + GRA analysis showing the benefits of minimization and prevention against the conventional approach. Savings generated by the reduction in external infrastructure costs are an important factor, as shown in this study. If other benefits, such as eco-system services, are also taken into consideration, the benefits from minimization and prevention could be even greater. However, at the moment these costs are difficult to estimate.

For the Cali expansion area (410,380 inhabitants), the best alternatives for both scenarios are B, C, G, and H. Two clearly differentiated sets of result from prioritizing, namely B and C, have a higher ranking, and G and H have a slightly lower ranking. These sets of alternatives share the same uses and sources of water, but differ in the excreta flushing equipment. With regard to the CBA, the ranking for different minimization and prevention alternatives remains the same for both scenarios (1 and 2); the cost-benefit ratios for scenario 2 are less favourable, mainly because the avoided costs due to the infrastructure of the sewer system and wastewater treatment plant are not taken into account.

In general, with the minimization and prevention, the water demand decreases according to the percentage of households that implement it. The urban model with the highest percentage of multi-family dwellings is the one that generates the lowest wastewater per capita. In this type of urban development, grey water is used for irrigation purposes and cleaning of communal areas, while single-family dwellings generate grey water in excess, due to the fact that this type of household does not have communal areas.

Water-saving devices have been introduced in the market only relatively recently (15-20 years ago); therefore, and also due to the limited market at the moment, these are still relatively expensive in the study area. It is expected that the cost of these devices will come down significantly in time and when they are applied more widely. It could change the results of this study. Additionally, the inclusion of a hypothetical advanced scenario of minimization that considers e.g. dry sanitation and vacuum systems would change the entire urban landscape of technology alternatives.

3.5 Conclusions

The identification and validation of the advantages of 3-SSA can be studied for each step, considering the possible borders between each. For the first step, minimization and prevention, in the case of households, this can be achieved through different combinations of low consumption devices, use of grey water and rainwater harvesting. The best alternatives are selected considering multiple criteria: technical, social, environmental and economic. These alternatives have advantage compared to conventional approach (toilet 6 L per flush and drinking water for all uses) in terms of cost-benefit analysis, when in this comparison are considered: additional costs to implement of prevention and minimization strategies, and the benefits (avoided costs) obtained as a result of implementing these strategies.

According to the AHP and GRA processes, the best minimization alternative for Cali's expansion zone corresponds to Alternative **C** (WC dual flush; grey water and rainwater harvesting), for both scenarios 1 and 2. Alternative C is the best solution after CBA of the conventional solution and the minimization and prevention alternatives B, C (WC dual flush) and G, H (WC 2.3 L). This is because high efficiency WC equipment (2.3 L) is still relatively expensive in the local market.

Minimization and prevention alternatives B, C, G and H are the best alternatives according to the AHP and GRA results (scenarios 1 and 2), independent of the type of percentage distribution of single and multi-family households, and the percentage of households implementing minimization and prevention alternatives.

The CBA comparing the B, C, G, and H alternatives and the conventional approach determines scenario 1 as the alternative with the best cost-benefit ratio, when considering the reduction in water supply costs for households and savings in the water supply, sewage and wastewater treatment plant infrastructure. In both cases, scenarios 1 and 2, the alternatives with the highest ranking in the application of AHP and GRA, compete in terms of a lower net present value NPV in comparison with the conventional approach.

Minimization and prevention alternatives become viable when the percentage of multi-family households using such alternatives is increased. For the study area, the minimization and prevention alternatives are viable ($NPV_{Benefit}/NPV_{cost} > 1.0$) if these are implemented in more than 20% of households.

In urban models with a greater number of single-family households, the most feasible alternative is B, while in the event that the model has a larger number of multi-family households, the best alternative is C. The urban models generating the greatest benefits are those corresponding to 100% single-family homes (alternatives B, G, H) and to 100% multi-family homes (alternative C).

3.6 References

Bos, J.J., Gijzen, H. J., Hilderink, H. B. M., Moussa, M., Niessen, L. W. and de Ruyter-van Steveninck, E. D. (2004). Quick Scan Health Benefits and Costs of Water Supply and Sanitation. Netherlands Environmental Assessment Agency (RIVM), IMTA - UNESCO-IHE, in consultation with WHO, the Netherlands.

Brent, R. (2006).   Applied cost-benefit analysis, Second Edition Ed., Edward Elgar Publishing limited, Cheltenham, UK.

Cardona, M.M. (2007). Minimización de Residuos: una política de gestión ambiental empresarial. *Producción más Limpia* 1(2), 46-57. (In Spanish)

Daigger, G. (2009). Evolving Urban Water and Residuals Management Paradigms: Water Reclamation and Reuse, Decentralization, and Resource Recovery. *Water Environment Research*, 81(8), 809-823.

Deng, J. L. (1982). Control problems of grey systems. System & Control Letters 1(5), 288-294.

Departamento Administrativo de Planeación Municipal (2008). Cali en cifras. Alcaldía de Santiago de Cali, ed., Impresora Feriva S.A., Cali. (In Spanish)

El Espectador (2009). ABC para el ahorro de agua. Recomendaciones para un efectivo ahorro de agua potable. In: El Espectador, diciembre 29 de 2009, Bogotá, Colombia. (In Spanish)

EMCALI and Hidroccidente S.A. (2006). Estudio de Alternativas de Dotación de los Servicios Públicos de Acueducto, Alcantarillado y Complementario de Alcantarillado en la Zona de Expansión de la Ciudad de Cali Denominada 'Corredor Cali-Jamundí'. Cali, Colombia. (In Spanish)

Galvis A., Guerrero J. E., Saldarriaga G. and Buitrago C.A. (2006). Proyecto Selección de Tecnología para el Control de Contaminación por Aguas Residuales Domésticas. Revista ACODAL Año 49, N° 214, Bogotá, Colombia. (In Spanish)

Gijzen, H.J. (2006). The role of natural systems in urban water management in the City of the Future – a 3 step strategic approach. *Ecohydrology and Hydrobiology* 6(1-4), 115-122.

Ghisi, E., and Mengotti de Oliveira, S. (2007). Potential for potable water savings by combining the use of rainwater and greywater in houses in southern Brazil. *Building and Environment*, 42, 1731-1742.

Hajkowicz, S. and Collins K. (2007). A review of multiple criteria analysis for water resource planning and management. *Water Resources Management*, 21, 1553–1566

Hajkowicz, S. and Higgins A. (2008). A comparison of multiple criteria analysis techniques for water resources management. *Operational Research* 184, 255–265

Kahraman, C. (2008). Fuzzy Multi-Criteria Decision Making, Theory and Applications with Recent Developments, Springer Optimization and Its Applications vol. 16.

Liu, S., Butler, D., Memon, F. A., Makropoulos, C., Avery, L., and Jefferson, B. (2010). Impacts of residence time during storage on potential of water saving for grey water recycling system. *Water Research*, 44(1), 267-277.

Liu Q., Singh V. P. and Xiang H. (2005). Plot Erosion Model Using Grey Relational Analysis Method, *Hydrologic Engineering ASCE* 10(4), 288-294.

Makropoulos, C.K., Natsis, K., Liu, S., Mittas, K. and Butler, D. (2008). Decision support sustainable option selection in integrated urban water management. *Environmental Modelling & Software,* 23 (12), 1448-1460.

Mejía, F.J., Isaza, P.A., Aguirre, S., and Saldarriaga, C.A. (2004). Reutilización de aguas domésticas. In: XVI Seminario Nacional de Hidráulica e Hidrología, Armenia, Colombia. (In Spanish).

Ministerio de Ambiente, Vivienda y Desarrollo Territorial (2010). Resolución No. 509 de 2010. Comisión Reguladora de Agua Potable y Saneamiento Básico, Bogotá. (In Spanish).

Miranda, J.J. (2000). Gestión de proyectos, identificación, formulación, evaluación financiera, económica, social, ambiental, Banco Mundial y MM Editores Cuarta Ed., Colombia. (In Spanish).

Mutikanga, H.E., Sharma, S.K. and Vairavamoorthy K. (2011). Multi-criteria Decision Analysis: A Strategic Planning Tool for Water Loss Management. *Water Resources Management* 25, 3947-3969.

Navarro, I. (2007). Determinación de la viabilidad técnica y económica de un sistema de almacenamiento temporal de aguas de drenaje urbano para la ciudad de Bogotá, Universidad de los Andes, Bogotá. (In Spanish).

Nhapi, I., and Gijzen, H.J. (2005). A 3-step strategic approach to sustainable wastewater management. *Water SA*, 31(1), 133-140.

OPS (2008). Saneamiento en Latinoamérica y en el Caribe. Report produced by Organización Panamericana de la Salud, Área de Desarrollo Sostenible y Salud Ambiental, Washington D. C., USA.

Ottoson, J., and Stenström, T.A. (2003). Faecal contamination of grey water and associated microbial risks. *Water Research*, 37(3), 645-655.

Pimienta, R. (2000) Encuestas probabilísticas vs. no probabilísticas. In: Política y cultura, Universidad Autónoma Metropolitana-Xochimilco, México. (In Spanish).

Romero, C. (1997). Análisis de las decisiones multicriterio, Isdefe, Ingeniería de Sistemas. Madrid, Spain. (In Spanish).

Saaty, T.L. (1990) How to make a decision: The Analytic Hierarchy Process. *European Journal of Operational Research* 48 (1990) pp. 9-26.

Saaty, T.L. (2008). Relative measurement and its generalization in decision making why pairwise comparisons are central in mathematics for the measurement of intangible factors the analytic hierarchy/network process. *Statistics and operation and Research* 102(2), 251-318.

Sanchez, G. de las N. (2003). Técnicas participativas para la planeación: Procesos breves de intervención, Fundación ICA, Mexico D.F. (In Spanish).

Siebel, M. A., and Gijzen, H.J. (2002). Application of Cleaner Production concepts in Urban Water Management. In: Environmental Technology and Management Seminar, Institute Technology Bandung/Environmental Engineering Department, Bandung, Indonesia.

Siebel, M.A., Gijzen H.J. (2003). Application of cleaner production concepts in urban water management. In: Conferencia Internacional Usos Múltiples del Agua: Para la Vida y el Desarrollo Sostenible. Universidad del Valle, Cali, Colombia.

Sierra, J. F. (2006). Tratamiento y reutilización de aguas grises en proyectos de vivienda de interés social a partir de humedales artificiales, Universidad de los Andes, Bogotá, Colombia.

Tosun, N. and Pihtili H. (2010). Grey relational analysis of performance characteristics in MQL milling of 7075 Al alloy. *Advance Manufacturing Technology* 46 (5-8), 509–515.

UNEP/GPA - UNESCO-IHE (2004). Improving Municipal Wastewater Management in Coastal Cities. A Training Manual for Practitioners. Published by Training GPA, The Hague, the Netherlands.

Veenstra, S., Alaerts, G.J. and Bijlsma, M. (1997). Technology Selection. Water Pollution Control – A guide to the use of water quality management principles. Edited by Richard Helmer and Ivanildo Hespanhol. Published on behalf of WHO/UNEP by F & FN Spon, London, UK.

Velasquez, J.S. (2009). Estimación de la demanda de agua urbana residencial: factores que la afectan, conservación del recurso y planteamiento metodológico desde el ordenamiento territorial y las medidas de conservación, Universidad Nacional de Colombia, Medellín-Colombia. (In Spanish).

WHO (2004). Guidelines for Drinking-Water Quality. World Health Organization. Third Edition. Geneva, Switzerland.

Zambrano, D.A. (2012) Minimización y Prevención Como Estrategia Para el Control de la Contaminación por Aguas Residuales Municipales en la Zona de Expansión de Cali. MSc Thesis, Universidad del Valle, Cali, Colombia.

Zein, M.L.E. (2006). Integrated Decision Support System for Wastewater Management: Case Study Egypt, Master of Science, UNESCO-IHE Institute for Water Education, Delft, the Netherlands.

Zeng, G., Jiang, R., Huang, G., Xu, M. and Li, J. (2007). Optimization of wastewater treatment alternative selection by hierarchy grey relational analysis. *Environmental Management*, 82(2), 250-259.

Chapter 4

Conceptual framework to select urban drainage system technology based on the Three-Step Strategic Approach

Source: CVC photo file

This chapter is based on:
Galvis. A., Montaña. F., Zambrano. D., Van der Steen, N.P. and Gijzen, H.J. (2014) *A multi-criteria model to select urban drainage system technology to minimize impacts on receiving waters*, presented in 13th International Conference on Urban Drainage, Sarawak, Malaysia, 7-12 September 2014.

Abstract

A new Conceptual framework for technology selection for the collection and transport of wastewater and stormwater in urban areas is proposed. The CF includes a multi-criteria analysis, considering technical, environmental, operational, maintenance, cost and institutional aspects. The model considers: 1. Pollution prevention and waste minimization (Step 1 of the Three-Step Strategic Approach); 2. Sustainable drainage system selection; 3. Feasibility of surface drainage; 4. The selection of combined or separate sewers; 5. Further selection of sewer type. The Conceptual framework uses a relatively small number of criteria and the information requirements are limited. For the case study Las Vegas (Cali, Colombia), local institutions planned for a traditional separate sewer. The results obtained with the model were different: erosion control and watershed maintenance; detention tanks; combined sewer. Application of the Conceptual framework in Latin America can contribute to: 1) improving urban drainage planning, 2) considering more technological options 3) improving selection of urban drainage technology.

4.1 Introduction

The impact on water resources of municipal wastewater discharges has become a critical and ever-growing environmental and public health problem. In order to be able to efficiently address this problem, it is important to adopt an integrated approach that includes a reduction of contamination at source. In this context, the conceptual 'Three Step Strategic Approach' (3-SSA) was developed previously, consisting of: 1) Pollution prevention and waste minimization, 2) Treatment for reuse and 3) Enhanced natural self-purification (Gijzen, 2006). This article is primarily about Step 1, with emphasis on the relationships within the urban drainage system, such as the interactions between the source of pollution (households and impervious areas) and the wastewater treatment plant (WWTP). Pollution prevention could be achieved by other options than the traditional combined or separate sewers, depending on the local conditions.

Although there is no general agreement on which of the two traditional systems (combined or separate sewers) is better (De Toffol *et al.*, 2007), the trends over the last few decades have increasingly been to implement separate systems. Under all conditions it is necessary to evaluate the local conditions before making a decision (Carleton, 1990; Giraldo, 2000; Brombach *et al.*, 2005; Stanko, 2009; Schaarup-Jensen *et al.*, 2011). Most of the existing drainage systems have one or more of the following problems: negative impacts on receiving water by the storm water runoff discharges, runoff pollution, dilution of influent to the WWTP, discharges from combined sewer overflows (CSOs) or illegal connections to sanitary sewer systems (Marsalek *et al.*, 2006). Different types of technologies are available, including for example small diameter sewers and simplified sewers, which have been applied successfully in several countries in Latin America. A recent development for storm water management is the application of so-called sustainable urban drainage systems (SUDS). These are aimed at reproducing, as close as possible, the natural water cycle as it existed prior to urbanization. SUDS maximize the opportunities and benefits that can be achieved from storm water management (Fletcher *et al.*, 2015). On the other hand, the WSUD (Water-Sensitive Urban Design) concept considers the urban water cycle, integrating water supply, wastewater management, urban planning, land-scape aspects and resilience of cities (Abbott *et al.*, 2013). These strategies seek to obtain sustainable and resilient systems to respond to current needs and to address the emerging threats of the 21st century (Butler *et al.*, 2014). However, despite these conceptual advances, in practice they are still not used in water planning (Ahern, 2011). In Latin America, in the big cities the options for separate or combined sewerage are analysed (Giraldo, 2000), and technology selection methodologies have been developed focused on wastewater treatment (Von Sperling, 2005; Noyola *et al.*, 2013) and on-site sanitation (CEPIS, 2002).

In this context, the technology selection for planning the collection and transport of runoff and wastewater in urban areas is a complex issue. It includes the analysis of environmental, social, technical and economic characteristics of each case, available technologies and the interaction between the sewer, WWTP and receiving water body. This paper presents the development and application of a conceptual framework CF for technology selection for urban drainage. The CF is focused in collection and transport of runoff and wastewater and not include technology selection for WWTP. The CF can be applied in new urban areas and in expansion areas of existing cities. The CF is based on the 3-SSA and it was developed for urban conditions in the cities of the Upper Cauca river basin (Colombia). These conditions are representative of urban drainage systems in Latin American cities. The flow chart of CF was designed to help decision makers in the selection of urban drainage strategies with the purpose of optimizing the effects of investments. However, the objective is not only economic but also to reach best management practices, including protection against flooding risks and to make a contribution to proper water resources management.

4.2 Methods

4.2.1 Definition of thematic blocks

The CF for technology selection was developed in five blocks. The first thematic block aims at pollution prevention and waste minimization (Block 1) and is a mandatory step prior to the other blocks. Two other blocks concern the selection of SUDS (Block 2) and the management of surface runoff through roads and gutters, without the need to install underground pipelines (Block 3). Block 4 concerns the choice between a separate sewer and a combined sewer. If the separate sewer system is chosen, the selection process continues to Block 5, where the selection of the wastewater sewer system is made, considering three options: a conventional sewer, small diameter sewer and simplified sewer.

Block 1: Pollution prevention and waste minimization
Pollution prevention and waste minimization practices are designed to reduce or eliminate the pollutants at source to prevent them from entering into the sewer system. The CF considers erosion control and watershed maintenance; comprehensive management of solid waste; cleaning of roads; management of household chemicals and efficient use of water. The user can switch these options on or off and the CF considers their effect on the size and costs of the WWTP.

Block 2: SUDS selection
To control water quantity and to improve the water quality of urban runoff through infiltration and storage devices, the user may choose the following SUDS options: permeable paving, an infiltration tank, a detention tank, a retention pond, and constructed wetland. According to Azzout *et al.* (1995), Barraud *et al.* (1999), and Ellis *et al.* (2008), decision making to select SUDS is divided into two stages: screening and selection using a multi-criteria analysis. Selection criteria were defined based on the developed models by Veldkamp *et al.* (1997), Brito (2006), Martin *et al.* (2007), Ellis *et al.* (2008), and literature reviews by Madge (2004); Castro *et al.* (2005), Woods-Ballard *et al.* (2007) and Perales (2008). Considering urban conditions in the cities of the Upper Cauca river basin, in Colombia, performance scores of the alternatives

for each indicator were obtained from literature reviews (Madge, 2004; Woods-Ballard *et al.*, 2007; Ellis *et al.*, 2008). Multi-criteria analysis was used for SUDS selection.

Block 3: Assessing the surface drainage feasibility
In this block the CF evaluates the possibility of using the hydraulic capacity of roads and ditches to drain (fully or partially) the surface flow that has not been captured by SUDS. By partial discharges of runoff to smaller networks of natural and artificial streams available in the study area, it is possible to avoid large volumes of runoff reaching the pipes of storm water or combined sewers, which allows a reduction in the pipe size and reduced associated costs. Criteria based on Bolinaga (1979), Van Duijl (1992) and Butler and Davies (2011), were used.

Block 4: Combined or separate sewer system
Runoff that was not managed by SUDS or surface drainage (blocks 2 and 3) has to be collected and transported by a sewer system to a final disposal point. Considering that combined and storm water sewers are technologies for this purpose, this block uses a multi-criteria framework to select the best technology for a given context. The identification of criteria was based on Van Duijl (1992), Meirlaen (2002), Butler and Davies (2011), Brombach *et al.* (2005) and De Toffol *et al.* (2007). The indicators used were characterized as follows.

Topography. This is characterized by the dominant slope of the drainage area and the diameter of the main sewer system. Sewer pipes require a specific slope (depending on the type of technology) to ensure self-cleaning conditions. In flat areas, pipes that require steeper slopes increase the volume of excavation and pumping height, which implies higher investment and O&M costs.

Wastewater pumping requirements. This was characterized by the percentage of wastewater that must be pumped. For this characterization, the following requirements of wastewater flow pumping were considered: 0% (by gravity); less than 20%, between 21% and 50% and above 50%.

First flush control. This indicator considers the percentage of drainage area managed by SUDS and type of SUDS selected in Block 2. The initial fraction of runoff is usually associated with a peak pollution concentration. In combined sewers, this fraction is led to WWTP, while in separate sewers this initial fraction is discharged directly into receiving water bodies or is infiltrated into the soil.

Dilution and self-purification capacity of receiving water body. This indicator considers the flow of the receiving water body during the dry season and the maximum flow of runoff. This maximum flow corresponds to the return period defined by local regulations, according to the size of the drainage area, its location and land use. Combined and separate sewer dis-charges during the rainy season generate an impact on the receiving water body quality. Based on the impacts of such discharges and water quality required to ensure water uses downstream from the discharge point, the decision may encourage the implementation of either technology.

O&M complexity. This indicator refers to the specific characteristics of the O&M required to control the flow and to avoid deterioration the runoff water quality during the lifetime of the system. It includes regular and occasional O&M and monitoring.

Illegal water connection control capacity. For this indicator three levels were defined: high, medium and low. This corresponds to the capacity of the municipal planning office to control incomplete development of settlements, reforms to existing homes and new urban developments and avoidance of illegal water connections to sewer systems.

Block 5: Selecting the type of sewer (wastewater management)
For wastewater management the options considered are combined sewers and separate sewer (septic tanks and small diameter pipes; simplified sewers, conventional sewers). Combined sewer technology was selected, when after applying Block 4 it turned out to be the best option for managing runoff, in whole or in part. For the other cases, a decision tree selected the best option. The following attributes were considered in this selection process: 1) Population density, 2) Social acceptance (in the case of small diameter sewer systems), 3) Interceptor tank maintenance guaranteed by the service provider, and 4) The availability of public spaces for the construction of interceptor tanks. If the implementation of small diameter sewers is not feasible, a selection between conventional sewers and simplified sewers is needed, considering the pavement width in order to install small diameter pipes and smaller manholes. The decision-making structure of Block 5 is supported by Mara (2005), Ministerio de Ambiente, Vivienda y Desarrollo Territorial (2010), U.S. EPA (2000), and WSP (2007).

4.2.2 Multi-criteria analysis

For the decision-making structure in Block 2 (SUDS selection) and Block 4 (Selection between combined and separate sewers), a multi-criteria analysis was applied, using the Weighted Sum Model - WSM (Zardari *et al.*, 2015). It is calculated using Equation 4.1.

$$P_{iWSM} = \sum_{j=1}^{n} a_{ij} W_j \qquad (4.1)$$

Where $P_{i\,WSM}$ is the weighted sum score for alternative *i*; *n* is the number of decision criteria, a_{ij} is the performance of alternative *i* in terms of the decision criterion *j*; and w_j is the weight of criterion *j*. The weighting of each criterion (w_j) represents its relative importance in the decision-making. The performance of the alternative (a_{ij}) corresponds to a rating score assigned when the alternative *i* is evaluated in terms of the criterion *j*. At the end of the calculations, the alternative with the highest score $P_{i\,WSM}$ is selected as the best option. The weighting of each criterion was obtained from 12 local experts. The performance of the alternatives for each criterion was rated based on the literature review and local experts. They belong to: sewer service companies, environmental authorities, consulting companies and academic research groups. Surveys carried out with experts included questions on: weighting of factors and criteria for SUDS selection and surface management of runoff. The experts rated the technologies in terms of type of sewer, the complexity of O & M and the institutional capacity to control the illegal connections.   For Block 3 (Assessing of surface drainage feasibility) and Block 5 (Selecting the type of sewerage), a decision tree (Lara *et al.*, 2004) was used.

4.2.3 Integration of the thematic blocks for the construction of the conceptual framework

At this stage, the thematic blocks were assembled in a diagram based on a methodology developed by Galvis *et al.* (2005). After defining the order of the blocks within the diagram, the integration was made taking into account the purpose of each block and their relationship with the next block.

4.2.4 Application of the conceptual framework for a case study

The developed CF for technology selection was applied to a case study for the area Las Vegas located in Cali, Colombia.

4.3 Results

4.3.1 Conceptual framework

The developed flow chart of CF is shown in the Figure 4.1. Specific results for blocks 2, 3 and 4 are presented below.

Block 2: SUDS selection
Decision making to select SUDS was divided into two stages: screening (Table 4.1) and selection using a multi-criteria analysis (Table 4.2).

Screening requires the following information: type of space available for construction of SUDS, soil infiltration rate, water table and catchment area. These data were compared with the characteristics required for preliminary selection of SUDS (Table 4.1) and alternatives that were not viable were discarded. Considering the type of required area for each SUDS, porous pavement preselection involves discarding other alternatives and vice versa, because for this CF they compete for different types of space. If the preliminary selection produces a single alternative, this is the chosen technology. However, if there are several viable alternatives, the multi-criteria analysis for SUDS selection is applied to make the choice.

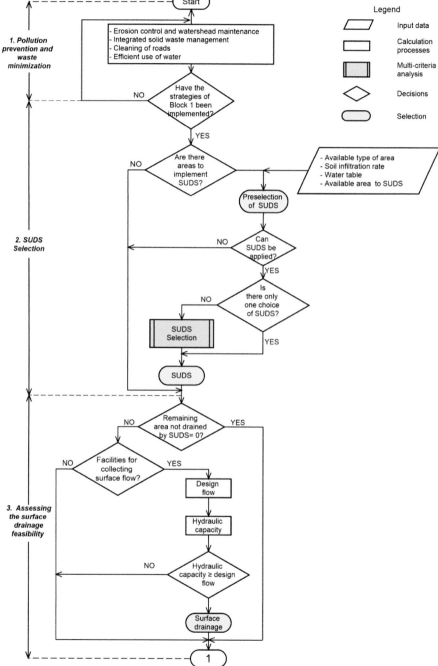

Figure 4.1 Flow-chart of technology selection for an urban drainage system, based on the 3-SSA

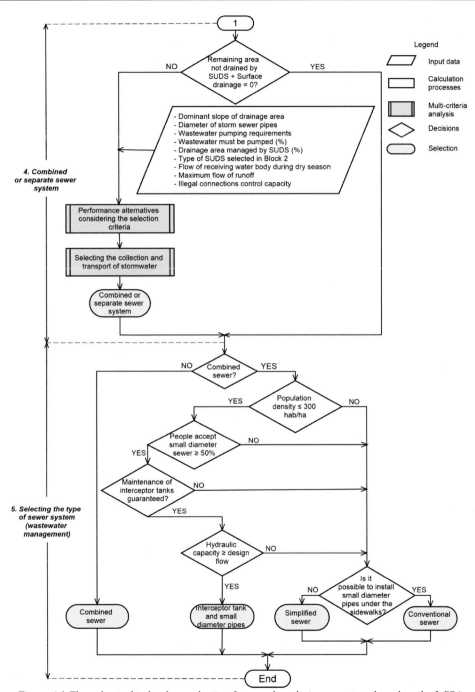

Figure 4.1 Flow-chart of technology selection for an urban drainage system, based on the 3-SSA
(cont.)

Table 4.1 Preliminary considerations for SUDS selection

Alternatives of SUDS	Characteristics for preliminary selection				
	Type of available area	Minimum required soil infiltration rate (mm/h)	Minimum water table depth[a] (m)	Area Available to implement SUDS[b] (ha)
Permeable paving	Parking or pavement	≥12	≥0.6	[c]
Infiltration tank	Park, green area or open space	12 - 76	≥1.2	≤5.0
Detention tank		[d]	≥1.0	≥4.0
Retention pond		[d]	≥1.0	≥6.0
Constructed wetland		[d]	≥1.0	≥8.0

Based on Madge (2004), Woods-Ballard *et al.* (2007) and Ellis *et al.* (2008)

[a] Depth from the bottom of the facility to the water table.

[b] Depending on the topography, may correspond to a part or the total area of urban development.

[c] Maximum ratio tributary area: area of porous pavement = 3:1

[d] This characteristic has no effect on the technology

Table 4.2 Weighted Sum Model (WSM) for SUDS selection

Category	Indicator	Weight w_j (%)	Performance of alternatives*a_{ij}			
			IT	DT	RP	CW
Technical	Hydraulic control to reduce peak flow	18.9	4	5	5	4
	Improvement of runoff water quality	11.3	5	2	2	4
	Potential for reuse of runoff water	7.3	0	0	5	3
O&M	Requirements for skills and materials	21.5	4	3	2	1
Urban	Aesthetic and amenity	14.5	3	4	5	5
Costs	Investment costs	13.7	5	3	3	2
	O&M costs	12.8	2	5	2	1
	P_{WSM}		3.56	3.45	3.36	2.77

IT: infiltration tank; DT: detention tank; RP: retention pond; CW: constructed wetland
*Performance as evaluated by experts from study area: 5: very high, 4: high, 3: medium, 2: low 1: very low. The assigned value has a direct relationship with the benefit to the urban drainage system. For example, if the lower cost strategy was the best option, then it would have the highest rating (5). Reference: Madge (2004) and Selvakumar (2004)

Based on the decision elements presented in the Table 4.2 and using Equation 4.1, the weighted sum score of each alternative (Pi WSM) was calculated. The technology with the highest score was selected. The Table 4.2 does not include the porous pavement because, as mentioned above, its preliminary selection implies discarding other alternatives.

Block 3: Assessing the surface drainage feasibility
Two conditions for decision-making were identified: 1) the availability of a smaller stream network in the area, and 2) the hydraulic capacity of the roads is larger than or equal to the rate of runoff. If both attributes are satisfied, the implementation of surface drainage will be viable. Otherwise, the selection process continues in Block 4 to choose the type of sewer system (combined or separate) for runoff collection.

Block 4: Combined or separate sewer system
Four criteria and six indicators (with their corresponding weights w_j) were used in Block 4. They were: 1) technical (topography: 19%; wastewater pumping requirements, by gravity or by pumping: 15.5%); 2) receiving water impacts (first flush control: 11.5%; dilution and self-purification capacity of receiving water body: 17%); 3) operation and maintenance (O&M complexity: 18.5%), and 4) institutional (illegal water connections control capacity: 18.5%).

Technology selection took place in two steps: (1) Assigning the performance scores of the alternatives based on the selection indicators and (2) Calculation of the weighted sum score (Pi WSM) of each alternative. In the first step, we should provide data about the context conditions and from them the performance of the alternatives regarding the selection criteria (a_{ij}) is obtained (see Table 4.3).

Table 4.3 Performance of alternatives in the case study area corresponding to selection indicators used in multi-criteria analysis

Indicators	Characteristics (local conditions)	Performance*a_{ij}	
		CS	SS
Wastewater pumping requirements	Gravity	4	3
	Pumping $\leq$ 20%	3	3
	Pumping 21 – 50%	2	4
	Pumping> 50%	1	4
First flush control	Drainage area managed by SUDS = 0%[a]	4	2
Dilution and self-purification capacity of receiving water body	Dilution factor[b] $\geq$ 40	4	4
	Dilution factor[b]< 40	2	3
O&M complexity	General[c]	2	3
Ability to control illegal connections	High	5	5
	Medium	5	3
	Low	5	1

CS: Combined sewer; SS: Storm water sewer
Performance: 5: very high, 4: high, 3: medium, 2: low, 1: very low
[a] If drainage area managed by SUDS is different to 0%, use Table 4.
[b] Dilution factor: relationship between the flow of the receiving water body during the dry season and maximum flow of runoff.
[c] General conditions of municipalities that belong to the study area.

If in Block 2 any type of SUDS was selected, then Table 4.4 is used to obtain the performance of the alternative to control the impact of the first flush. The scores presented in the Table 4.3 are based on Bertrand-Krajewski *et al.*, (1998), Barco *et al.*,(2008) and Ellis *et al.*, (2008).

4.3.2 Case study: Las Vegas (Cali), Upper-Cauca river basin, Colombia

This area is planned for a population of 15,000 inhabitants. Las Vegas consists of 59 ha, of which 84% is impervious (roads, car parking, pavements and roofs) and 16% is greenery. This urbanization is expected to have its own wastewater treatment plant. The sewer service company of Cali city has selected a separate sewer. In comparison, below are the results of applying the CF for technology selection.

Block 1: Pollution prevention and waste minimization
According to the particular conditions of this case study it is possible to implement: erosion control and watershed maintenance. This includes practices that help keep rainwater in the soil and maintain sediments in the site, e.g. the conservation of vegetation covers in parks, open spaces and upper basin and land use planning to reduce the area of impermeable surfaces and to prevent erosion and deforestation of the basin.

Table 4.4 Alternative performance according to ability to control the impact of the first flush

SUDS selected in Block 2	Drainage area managed by SUDS			
	< 20%		≥ 20%	
	CS	SS	CS	SS
Infiltration tank	4	3	5	4
Detention tank	3	2	4	3
Retention pond	3	2	4	3
Constructed wetland	4	3	5	4
Permeable paving	4	3	5	4

CS: combined sewer; SS: storm sewer
Performance: 5: very high, 4: high, 3: medium, 2: low, 1: very low

Block 2: SUDS selection
For the implementation of SUDS, two areas (9.5 ha and 12.4 ha) were identified, with the following characteristics: infiltration rate: 61.2 mm/h, water table depth: 3 m. Three alternatives were pre-selected: a detention tank, retention pond and constructed wetland. The detention tank was selected due to the best performance compared with the other two options. The runoff generated from 21.9 ha (37% of total area) would be evacuated to two detention tanks.

Block 3: Assessing the surface drainage feasibility

It was not feasible to implement surface drainage, since the area does not have facilities for collecting surface flow.

Block 4: Combined or separate sewerage system
Combined sewerage was selected for the 37.1 ha that was not connected to the detention tank selected in Block 2. The data used for making this decision are presented in the Table 4.5. The Table 4.6 shows the weighted sum score of the alternatives considered in this block ($P_{i\ WSM}$).

Block 5: Selecting the type of sewer system
The combined sewer in Block 4 was selected to drain part of the runoff, and also selected to transport wastewater in the urban development.

Table 4.5 Indicators for technology selection in Block 4 in Las Vegas, Cali, Colombia

Indicators	Value
Dominant slope of drainage area	0.003-0.005 m/m
Storm sewer pipe diameter	600-800 mm
Wastewater pumping requirements	0%
Drainage area managed by SUDS	37%
SUDS selected in Block 2	Detention tank
Receiving water body discharge during the dry season	0.513 m³/s
Maximum flow of runoff	1.5 m³/s
Ability to control illegal connections to sanitary sewer system	Medium

Table 4.6 $P_{i\ WSM}$ of combined and storm water sewerage in Las Vegas, Cali, Colombia

Indicators	Weight (%) w_j	Performance a_{ij} CS	SS	Partial score $a_{ij} \times w_j$ CS	SS
Topography	19.0	5	5	95	95
Wastewater pumping requirements	15.5	4	3	62	46.5
First flush control	11.5	4	3	46	34.5
Dilution and self-purification of receiving water body	17.0	2	3	34	51
O&M complexity	18.5	2	3	37	55.5
Ability to control illegal connections to sanitary sewer	18.5	5	3	92.5	55.5
$P_{i\ WSM}$				366.5	338

CS: combined sewer; SS: storm water sewer.
Performance: 5: very high, 4: high, 3: medium, 2: low, 1: very low

4.4 Discussion

The concept of integrated water management in urban areas emerged in the 60s (Fletcher *et al.*, 2015) but whereas in recent decades emphasis on the need for this integrated management has increased, it has not been common practice (Vanrolleghem *et al.*, 2005). Nevertheless, the need to consider urban drainage as a complex system, instead of only as an infrastructure, is recognized. This involves the visions of different types of decision makers (Fratini *et al.*, 2012). However, in practice, environmental authorities, service providers and consulting companies continue to operate with the same strategies of more than 50 years ago. In Colombia, and in general in Latin America, it is common practice to define a priori that it must be a separate sewer, although in practice many separate sewers are operating like combined sewers, by malfunctioning, illegal connections, informal settlements, etc.

Adopting the integrated water management concept, the process of urban drainage technology selection depends on local conditions and the interaction between the sewer system, the WWTP and the self-purification capacity of the receiving water body. In this context and the framework of 3-SSA (Gijzen, 2006), the CF to select urban drainage system proposed in this paper was developed. Although the focus of the CF is on Step 1 (Prevention of pollution and minimization of waste), it also includes Step 2 (Treatment for reuse) and Step 3 (Improved natural self-purification), when the CF considers the sewer systems as an integral part of the urban water cycle. The results obtained in this research show the importance of explicitly including the sewage system in 3-SSA and consider both point pollution and diffuse pollution control to contribute to sustainable water resource management.

The CF was applied in the urban expansion area Las Vegas, in Cali, Colombia. For this urbanization, the local institutions considered only a traditional solution with a separate sewer. On the other hand, the results obtained with the CF includes: minimization and prevention strategies (Block 1); SUDS (Block 2) and combined sewers (blocks 4 and 5).

4.5 Conclusions

Technology selection for urban drainage systems plays an important role in the efficient management of runoff and wastewater. It is a complex decision involving different criteria: environmental, social, technical, economic and institutional. This process should include several technological options. Thus, the multi-criteria methodology allows the use of knowledge of local experts in the design and construction of this type of decision process. The developed CF was based on the Three Step Strategic Approach (3-SSA). It used a relatively small number of criteria. This is an advantage for its potential users. Additionally, these criteria and the information requirements are easily recognized by both decision-makers and designers in Colombia and the Latin America contexts.

For the proposed CF, technology selection of urban drainage should consider the following sequence: 1. Pollution prevention and waste minimization at different levels (Step 1 of the 3 SSA); 2. SUDS selection; 3. Assessment of the surface drainage feasibility; 4. The choice between combined or separate sewer systems; 5. Selection of the type of sanitary sewer, including: small diameter sewers with interceptor tanks, simplified sewer and separate sewer.

For the case study Las Vegas in Cali, Colombia, local institutions considered only a traditional solution with a separate sewer. The results obtained with the CF are as follows: erosion control and watershed maintenance (Block 1); SUDS (detention tank) to handle 37% of runoff in the drainage area (Block 2) and combined sewer (blocks 4 and 5). For Block 4, the scores for combined and separate sewers are similar. However, the limited capacity of the local institutions to control illegal connections to sewer systems confirmed the decision to select a combined sewer.

The application of the CF in Colombia and other Latin American countries can contribute to: 1) improving urban drainage systems planning, 2) considering the broader technological offer, beyond traditional (a separate sewer and a combined sewer); 3) improving the selection of urban drainage system technology.

The results of this research can be applied to new case studies. On the other hand, software development could facilitate the CF application and therefore make the validation process much more efficient.

4.6 References

Abbott, J., Davies, P., Simkins, P., Morgan, C., Levin, D. and Robinson, P. (2013). Creating water sensitive places - scoping the potential for Water Sensitive Urban Design in the UK. CIRIA, London, UK.

Ahern, J. (2011). From fail-safe to safe-to-fail: Sustainability and resilience in the new urban world. *Landscape and Urban Planning,* 100 (4), 341-343.

Azzout, Y., Barraud, S., Cres, F.N. and Alfakih, E. (1995). Decision aids for alternative techniques in urban storm management. *Water Science & Technology,* 32 (1), 41-48.

Barco, J., Papiri, S. and Stenstrom, M.K. (2008). First flush in a combined sewer system. *Chemosphere,* 71 (5), 827-833.

Barraud, S., Azzout, Y., Cres, F.N. and Chocat, B. (1999). Selection aid of alternative techniques in urban storm drainage - Proposition of an expert system. *Water Science & Technology,* 39 (4), 241-248

Bertrand-Krajewski, J.L., Chebbo, G. and Saget, A. (1998). Distribution of pollutant mass vs volume in stormwater discharges and the first flush phenomenon. *Water Research,* 32 (8), 2341-2356.

Bolinaga, J. (1979). *Drenaje Urbano,* Instituto Nacional de Obras Sanitarias. Caracas, Venezuela. (In Spanish).

Brito, D.S. (2006). Metodologia para seleção de alternativas de sistemas de drenagem. Dissertação de Mestrado, Departamento de Engenharia Civil e Ambiental, Universidade Brasilia, Brazil, 117. (In Portuguese)

Brombach, H., Weiss, G. and Fuchs, S. (2005). A new database on urban runoff pollution: comparison of separate and combined sewer systems. *Water Science and Technology,* 51 (2), 119-128.

Butler, D. and Davies, J.W. (2011). *Urban Drainage,* London, Third Edition published by Spon Press. ISBN 0-20384905.

Butler, D., Farmani, R., Fu, G., WARD, S., Diao, K. and Astaraie-Imani, M. (2014). A New Approach to Urban Water Management: Safe and Sure. *Procedia Engineering,* 89 (2014), 347-354.

Carleton, M.G. (1990). Separate and combined sewers. Experience in France and Australia. In Massing, H., Packman, J., Zuidema, F. (Eds.) *Hydrological Processes and Water Management in Urban Areas.* Wallingford, UK, International Association of Hydrological Sciences.

Castro, D., Rodríguez, J., Rodríguez, J. and Ballester, F. (2005). Sistemas Urbanos de Drenaje Sostenible (SUDS). *Interciencia,* 30 (005), 255-260. (In Spanish).

Centro Panamericano de Ingeniería Sanitaria y Ciencias del Ambiente CEPIS (2002) Algoritmo para la Selección de la Opción Tecnológica y Nivel de Servicio en Saneamiento. División de Salud y Ambiente Organización Panamericana de la Salud Oficina Sanitaria Panamericana. Lima, Perú. (In Sapnish).

De Toffol, S., Engelhard, C. and Rauch, W. (2007). Combined sewer system versus separate system – a comparison of ecological and economic performance indicators. *Water Science & Technology,* 55 (4), 255-264.

Ellis, J. B., Revitt, D. M. and Scholes, L. (2008). The DayWater Multi-Criteria Comparator. InThévenot, D. R. (Ed.) DayWater: An Adaptative Decision Support System for Urban Stormwater Management. London, IWA Publishing.

Fletcher, T.D., Shuster, W., Hunt, W.F., Ashley, R., Butler, D., Arthur, S., Trowsdale, S., Barraud, S., Semadeni-Davies, A., Bertrand-Krajewski, J.L., Mikkelsen, P.S., Rivard, G., Uhl, M., Dagenais, D. and Viklander, M. (2015). SUDS, LID, BMPs, WSUD and more – The evolution and application of terminology surrounding urban drainage. *Urban Water Journal* 12(7), 1-18.

Fratini, C.F., Geldof, G.D., Kluck, J. and Mikkelsen, P.S. (2012). Three Points Approach (3PA) for urban flood risk management: A tool to support climate change adaptation through transdisciplinarity and multifunctionality. *Urban Water Journal* 9(5), 317-331.

Galvis A., Cardona, D. A. and Bernal D.P. (2005). Modelo Conceptual de Selección de Tecnología para el Control de Contaminación por Aguas Residuales Domésticas en Localidades Colombianas Menores de 3000 Habitantes SELTAR. Proceedings IWA Agua 2005. Conferencia Internacional: De la Acción Local a las Metas Globales, November 2005, Cali, Colombia. (In Spanish).

Gijzen, H.J. (2006). The role of natural systems in urban water management in the City of the Future – a 3 step strategic approach. *Ecohydrology and Hydrobiology* 6 (1-4), pp. 115-122.

Giraldo, E. (2000). Combinar o Separar? Una discusión con un siglo de antigüedad y de gran actualidad para los bogotanos. *Revista de Ingeniería de la Universidad de los Andes,* 11, 21-30. (In Spanish).

Lara, J., Quijano, J., Riveros, D., Torres, A. and Forero, M. (2004). Utilización de sistemas expertos para la optimización de la toma de decisiones multicriterio. *XXIX Congreso Interamericano de Ingeniería Sanitaria y Ambiental.* San Juan, Puerto Rico. (In Spanish).

Madge, B. (2004). Effective Use of BMPs in Stormwater Management. In U.S. EPA (Ed.) The Use of Best Management Practices (BMPs) in Urban Watersheds. Washington D.C., United States Environmental Protection Agency, Office of Research and Development.

Mara, D.D. (2005) Sanitation for All in Periurban Areas? Only If We Use Simplified Sewerage. In: *Water Science & Technology: Water Supply* 5(6), 57-65. London: IWA Publishing.

Marsalek, J., Jiménez-Cisneros, B.E., Malmquist, P.A., Karamouz, M., Goldenfum, J., and Chocat B. (2006). *Urban Water Cycle Processes and Interactions,* Paris, IHP International Hydrological Programme.

Martin, C., Ruperd, Y. and Legret, M. (2007). Urban stormwater drainage management: The development of a multicriteria decision aid approach for best management practices. *European Journal of Operational Research,* 181 (1), 338-349.

Meirlaen, J. (2002). Immission based real-time control of the integrated urban wastewater system. PhD Thesis, Faculteit Landbouwkundige en Toegepaste Biologische Wetenschappen, Gent University, Belgium, 260.

Ministerio de Ambiente, Vivienda y Desarrollo Territorial (2010). Reglamento Técnico del Sector de Agua Potable y Saneamiento Básico: TÍTULO J. Alternativas tecnológicas en agua y saneamiento para el sector rural. Bogotá, D.C. Colombia, ISBN: 978-958-8491-42. (In Spanish).

Noyola, A., Morgan- Sagastume, J.M. and Guereca, L.P. (2013). Selección de Tecnología para el tratamiento de aguas residuales municipales. Universidad Autonoma de México. ISBN: 978-607-02-4822-1. (In Spanish).

Perales, S. (2008). Sistemas Urbanos de Drenaje Sostenible (SUDS). Expo Zaragoza 2008. Agua y Desarrollo Sostenible. 5ª Semana Temática de la Tribuna del Agua. Zaragoza, Spain. (In Spanish).

Schaarup-Jensen K., Rasmussen M.R. and Thorndahl S.L. (2011). The Effect of Converting Combined Sewers to Separate Sewers. 12th International Conference on Urban Drainage: proceedings: Porto Alegre/Brazil, 10-15 September 2011

Selvakumar, A. (2004). BPM Costs. In U.S.EPA (Ed.) The Use of Best Management Practices (BMPs) in Urban Watersheds. Washington D.C., United States Environmental Protection Agency, Office of Research and Development.

Stanko, S. (2009). Combined versus Separated Sewer System in Slovakia. *International Symposium on Water Management and Hydraulic Engineering.* Ohrid, Macedonia.

U.S. EPA (United States Environmental Protection Agency) (2000). Decentralized Systems Technology Fact Sheet Small Diameter Gravity Sewers. EPA 832-F-00-038. Washington D.C.

Van Duijl, L.A. (1992). Urban Drainage and Waste Water Collection, IHE Delft, The Netherlands.

Vanrolleghem, P.A., Benedetti, L. and Meirlaen, J. (2005). Modelling and real-time control of the integrated urban wastewater system, *Environmental Modelling & Software*, Vol. 20 (19), pp. 427

Veldkamp, R., Hermann, T., Colandini, V., Terwel, L. and Geldof, G. (1997). A decision network for urban water management. *Water Science & Technology,* 36 (8-9)**,** 111-115.

Von Sperling, M. (2005). *Introduction to water quality and sewage treatment*, 3th edition, Department of Sanitary and Environmental Engineering – Federal University of Mina Gerais, Belo Horizonte. (In Portuguese)

Woods-Ballard, B., Kellagher, R., Martin, P., Jefferies, C., Bray, R. and Shaffer, P. (2007). *The SUDS Manual,* London, UK.

WSP (Water and Sanitation Program World Bank) (2007). La ciudad y el saneamiento. Sistemas condominiales: Un enfoque diferente para los desagües sanitarios urbanos, Lima, Peru.

Zardari, N.H., Ahmed, K., Shirazi, S.M. and Yusop, Z.B. (2015). Weighting Methods and their Effects on Multi-Criteria Decision Making Model Outcomes in *Water Resources Management*. ISBN: 978-3-319-12585-5.

Chapter 5
Financial aspects of reclaimed wastewater irrigation in three sugarcane production areas in the Upper Cauca river Basin, Colombia

Source: photo file Cinara - Univalle

This chapter is based on:
Galvis. A., Jaramillo. M.F., Van der Steen, N.P. and Gijzen, H.J. (2018). Financial aspects of reclaimed wastewater irrigation in three sugarcane production areas in the Upper Cauca river Basin, Colombia. *Agricultural Water Management*, 209, 102-110; doi: 10.1016/j.agwat.2018.07.019

Abstract

Treated wastewater may be reused for crop irrigation. This contributes to recovery of water and nutrients, and at the same time it helps to reduce pollution discharge to receiving water bodies. Despite these advantages of reuse of treated wastewater, there is little experience of this in Colombia and Latin America. In part, this condition is explained by the lack of studies that show the potential of reuse comparing the traditional wastewater treatment options without reuse versus the options with reuse. In this research, the financial viability of reuse of treated wastewater for the irrigation of sugarcane crops in the Upper Cauca river basin in Colombia was studied. The study included three cases, with different characteristics of wastewater (BOD_5 between 164 and 233 mg/L), flows (between 369 and 7,600 L/s), rainfall levels (between 1,009 and 1459 mm/year) and irrigation requirements (0.34 and 1.08 L/s-ha). For both scenarios, the same baseline was considered. Cost-Benefit Analysis CBA was used to compare the options (with and without reuse of treated wastewater). Cost of initial investment and O&M were considered. Benefits were considered like avoid cost in use of fertilizers, reduction of taxes for water use and discharges directly to water bodies and investment and O&M costs of infrastructure for irrigation with groundwater. The results of the CBA and sensitivity analysis show that there are two key factors that influence financial viability of treated wastewater for sugarcane crop irrigation: 1) the water balance and irrigation requirements, and 2) costs corresponding to the management of wastewater for agricultural irrigation, including additional treatment (when it is required) and the infrastructure to bring the treated wastewater to crops. The financial viability of reuse in the study area was limited because the values of tax for wastewater discharges and water tariffs in Colombia do not correspond to the values they should have.

5.1 Introduction

The reuse of treated wastewater in agriculture corresponds to the subsequent use that is given in the irrigation of crops (Brega Filho and Mancuso, 2003). This puts demands on the municipal wastewater treatment technology to meet specific quality standards corresponding to the type of reuse defined (Asano *et al.*, 2007; U.S. EPA, 2012). A main objective of wastewater treatment is to reduce the environmental impact on receiving water bodies via pollution reduction. The reduction of this environmental impacts may be achieved when treated wastewater is reused in activities such as crop irrigation, industrial processes, cleansing or washing activities (Becerra *et al.*, 2015; Capra and Scicolone, 2007; Winpenny *et al.*, 2013). Besides reusing water, this approach will also promote the reuse and recovery of other resources such as nutrients and energy (Gijzen, 2006, 2001). Additionally, wastewater reuse, being an additional source of water, represents environmental benefits such as maintenance of critical water flows in sensitive ecosystems and recreational activities. Treated wastewater reuse is the second step The Three-Step Strategic Approach (3-SSA). It presents an integrated approach toward pollution and water quality management, consisting of: 1) minimisation/prevention, 2) treatment for reuse, and 3) planned discharge with stimulation of self-purification capacity of receiving waters (Gijzen, 2006; Galvis *et al.*, 2014). To ensure maximum impact and benefits, the three steps should be implemented together, preferably in a chronological order, and possible interventions under each step should be fully exhausted before moving on to the next step (Galvis *et al.*, 2018).

Agriculture is the main user of freshwater, accounting for over 70% of the total global freshwater withdrawal from rivers, lakes and aquifers (UNESCO - IHP, 2014; Winpenny *et al.*, 2013). For countries with low incomes (Gross National Income GNI < US$ 1005) or lower middle incomes ($US 1006 < GNI < $US 3955) this value corresponds to 82% (Amigos de la Tierra América Latina y el Caribe, 2016). On the other hand, wastewater is used in agriculture. However, in most cases effluent reuse in agriculture is done without any treatment, and this poses direct health risks. The Food and Agriculture Organization of the United Nations (FAO) reported that approximately 90% of total irrigation (with wastewater) is by untreated or partially treated wastewater (Winpenny *et al.*, 2013). Millions of hectares are irrigated with raw wastewater in regions such as China, Mexico and India (Jiménez and Asano, 2008). In addition it is estimated that at least 20 million hectares are irrigated in 50 countries with polluted water (Jimenez and Asano, 2004). The main limitation of these reuse practices is that sewage is generally not (sufficiently) treated before reuse, which introduces public health risks and environmental impacts. The challenge is to develop adequate treatment systems that produce biologically and public health safe effluents, but which preserve valuable components such as nutrients, which may replace fertilizers.

The potential of reuse technologies has been considered primarily following the WHO and the United States Environmental Protection Agency U.S. EPA guidelines (WHO, 2006; U.S. EPA, 2012). WHO recommends natural (or extensive) systems, which are more viable in developing countries (tropical and subtropical), in terms of operation and maintenance. Although protozoa and helminths are key parameters to be considered in the reuse applications, because of their impact on health, they are not considered in the regulations of some developed countries. In contrast, the technologies used in developed regions to treat wastewater for reuse have high removal efficiencies for other pathogens (Jimenez et al., 2010; Moscoso et al., 2002).

In a study of the financial viability of reuse, the following types of benefits can be considered: 1) Savings on water use, the use of treated effluents will reduce the use of freshwater resources for irrigation; 2) Savings for the reduced use of fertilizers. Effluent reuse improves soil fertility contributing organic matter and macronutrients (N, P, K), thus reducing the use of chemical fertilizers (Corcoman et al., 2010; Hespanhol, 2003; Winpenny et al., 2013); 3) Reduction in sewer tariffs and tax for wastewater discharges directly to water bodies. The reduction of effluent discharges contributes directly to the improvement of the water quality of the receiving water bodies (Bixio and Wintgens, 2006); 4) Converting Chemical Oxygen Demand COD into energy. Wastewater treatment can be accomplished in aerobic or anaerobic systems, but anaerobic systems appear to be more favourable because of energy recovery and cost-effectiveness (Gijzen, 2001); 5) Savings on infrastructure and its operation and maintenance (O&M) for irrigation when groundwater is used. With the wastewater reuse in agriculture, groundwater is preserved since agricultural reuse will contribute a percentage of its recharge with superior quality characteristics (Moscoso et al., 2002). In addition to agricultural reuse, infrastructure costs and pumping groundwater may be avoided (Cruz, 2015). This strategy therefore contributes to reducing freshwater use, closing nutrient cycles, reducing pollution discharges into receiving water bodies, and reducing infrastructure and O&M costs. As such, effluent reuse reduces the eutrophication of water bodies and costs in freshwater and the use of agrochemicals in farming (Candela et al., 2007).

Despite the advantages mentioned above of reuse of treated wastewater, there is little experience of this in Colombia and Latin America. In Colombia, only raw wastewater is used in agricultural irrigation. This situation has been generated mainly by: i) inadequate management of domestic wastewater, ii) undervaluation of wastewater as an alternative source of irrigation, iii) ignorance of the conceptual aspects for the implementation of reuse and iv) policies and regulations for the reuse management are not adequately articulated (Ministerio de Agricultura y Desarrollo Rural, 2011; Ministerio de Ambiente y Desarrollo Sustentable MADS, 2014). The assessment of the financial viability of reuse in agricultural irrigation can help stimulate reuse as a strategy for wastewater management. This assessment can be done using the cost benefit analysis CBA to compare the wastewater management considering reuse and without considering the reuse option. This study considers the local context regarding the type of reuse, the cost of raw water and local regulations. In this research, the financial viability of the reuse of treated wastewater in sugarcane crop irrigation in the Upper Cauca river basin in Colombia was studied. The research included three case studies, with different characteristics of wastewater, effluents flows from wastewater treatment, rainfall levels and irrigation requirements. For both scenarios (with and without reuse), the same baseline was considered and CBA was used to compare the two scenarios. A sensitivity analysis was performed for the irrigation requirements, water use fee rate and tax for wastewater discharges to water bodies (effluent charges). In this study an incremental analysis of CBA was employed (Boardman *et al.*, 2001; Harrison, 2010). It does not consider the common costs and benefits, such as health benefits, of the two scenarios that were compared.

5.2 Material and Methods

5.2.1 Study area

Upper Cauca river basin. The Cauca is the second most important river of Colombia and the main water source of the Colombian southwest. It has a longitude of 1,204 km with a basin of 59,074 km². The study area is the Upper Cauca river basin (Figure 5.1), in particular the corresponding to stretch La Balsa - Anacaro. La Balsa is 27.4 km (980.52 meters above sea level (m.a.s.l) and Anacaro 416.1 km (895.56 m.a.s.l.). The 0.0 km corresponds to Salvajina dam. This stretch of the Cauca River has an average width of 105 m. The depth can vary between 3.5 and 8.0 m. The longitudinal profile of the river shows a concave shape with a hydraulic slope, which oscillates between 1.5×10^{-4} m/m and 7.0×10^{-4} m/m (Ramirez *et al.*, 2010). The average annual rainfall varies between 938 mm (central sector) and 1882 mm (southern sector). There are two dry season periods: December - February and June - September. Rainy days per year vary between 100 days (central sector) and 133 days (northern sector) (Sandoval and Ramírez, 2007).

An important part of the sugarcane crops and the Colombian sugar industry are located in the flat area along the Upper Cauca river basin. In this flat area are the largest cities and therefore here where the largest amount of wastewater is generated. In the mountain area, there are coffee crops and associated industry. The Cauca River has been used for the last decades in fishing, recreation, power generation, riverbed matter extraction, domestic water supply, irrigation and industry. The Salvajina reservoir began operations in 1985 and it is part of a project aimed at improving flood control, water quality, self-purification capacity and power generation. The reservoir has a capacity of 270 MW. The reservoir operates with a minimum flow discharge of 60 m³/s and an average daily flow rate of 140 m³/s at the Juanchito station (Sandoval *et al.*, 2007). The Cauca River receives solid waste and wastewater discharges from industrial and domestic sectors, which is contributing to the decline in water quality. In the study area, there are currently 3.9 million inhabitants and the Cauca River receives, in the La Balsa - Anacaro stretch, approximately 140 ton/d of Biochemical Oxygen Demand (BOD_5).

Selecting the case studies. This study area was selected considering: 1) the existence of large sugarcane crops with potential to be irrigated with treated wastewater; 2) scarce surface water for agricultural irrigation during sometimes of the year; 3) location of cities with wastewater treatment systems (existing and / or projected); 4) availability of information on the characteristics of the sugarcane crop, due to the existence in the region of a sugarcane research centre (Cenicaña); 5) hydro-meteorological information availability through Cenicaña and the regional environmental authorities; 6) availability of information on wastewater management through the service provider companies and the regional environmental authorities. Three case studies of the Upper Cauca river basin were selected. Different sizes of wastewater treatment plants, different water quality in the influent to wastewater treatment plant and different water demands for irrigation were considered (Table 5.1). The three cases are located on the Cauca River bank at a distance of approximately 80 km. For all the cases sugarcane was the crop for irrigation. The average temperature was 24 ° C.

Wastewater treatment with reuse. The wastewater treatment alternatives considered in the three cases the national regulations for discharge of effluents to surface water bodies. In 2013 Colombia did not have a water quality regulation for the use of effluents of wastewater treatment plants. WHO and FAO guidelines were applied. In this case the use of treated wastewater for irrigation of crops commercially processed before human consumption was considered. For the effluent, the limiting conditions for reuse were: helminth eggs (HO) / liter <1.0/L (WHO, 2006); faecal coliforms< 10^3/100 mL; BOD <30 mg/L and TSS <30 mg /L (Pescod, 1992). Additionally, the treatment system requires compliance with agronomic quality guidelines (Ayers and Wescot, 1987).

Table 5.1 Characteristics of case studies. Reference year: 2013

Case studies	Population (inhabitants)	Flow of wastewater (L/s)	Existing wastewater treatment level	BOD influent (mg/L)	Average precipitation (mm/year)
1. Sewered area of Cali	2,060,000	7,600	Advanced primary	164	1,015
2. Expansion area of Cali	410,380	850	Untreated	226	1,459
3. Buga city	135,341	369	Untreated	233	1,009

5.2.2 Case studies

Case 1. **Sewered area of Cali.**

Without reuse. This case study was prepared based on information from EMCALI (2007a, 2007b, 2011) and Universidad del Valle and EMCALI (2008). The technology without reuse corresponded to: Advanced primary treatment (existing system) + Activated sludge, contact stabilization (projected). The effluent from the existing treatment plant (advanced primary treatment) discharged to the Cauca River 59% of the total contaminant load produced by Cali city. This value is explained because there is a part of the city of Cali that is not connected to this WWTP. The drainage area is located between 950 m.a.s.l. and 1,100 m.a.s.l. This area has a sewerage system that collects approximately 70% of the municipal wastewater of Cali city.

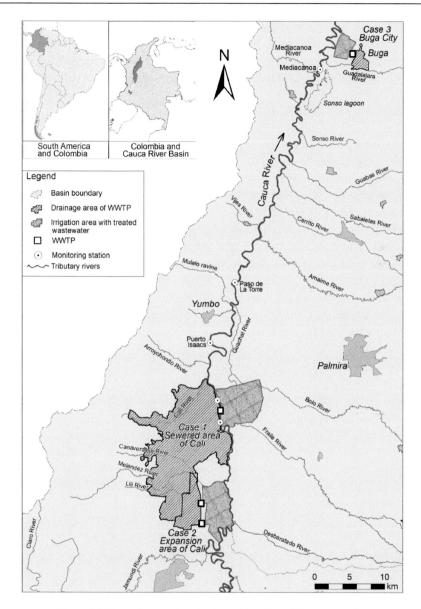

Figure 5.1 Upper Cauca river basin and location of the case studies

With reuse. This case includes the existing advanced primary treatment, a planned secondary treatment for agricultural reuse and the use of the effluent from the secondary treatment system for irrigation of sugarcane crops on the right bank of the Cauca River. Of the 7,600 L/s that the WWTP receives, 56% (4,274 L/s) is considered for treatment by the technological option without reuse - which corresponds to the existing situation and expansion plans, while 44% (3,326 L/s) is considered for treatment by the technological option with reuse: Preliminary and

Advanced Primary treatments (existing) + Upflow Anaerobic Sludge Blanket (UASB) + Facultative pond + Maturation pond. As the existing infrastructure is on the left bank and the area to be irrigated is on the right bank of the river, it will be necessary to build a viaduct. The costs of this viaduct were obtained from Universidad del Valle and EMCALI (2008).

Case 2. **Expansion area of Cali.**

Without reuse. This case study was prepared based on information from EMCALI and Consorcio Ingesam-Hidroccidente (2009); EMCALI and Hidroccidente (2006) and Gaviano *et al.* (2009). The technology without reuse (two lines, each of 425 L/s) corresponds to: Preliminary treatment + High rate anaerobic lagoon (Peña, 2002) + Facultative pond. The expansion area (1,358 ha) is located to the south of Cali, in the alluvial plain of the Pance and Cauca rivers. The slopes are between 0 and 2 degrees and with elevations between 955 m.a.s.l and 1,030 m.a.s.l. Land use is extensive livestock farming and sugarcane cultivation, as well as some developed areas with institutional, recreational and rural housing.

With reuse. Technology with agricultural reuse was based on Gaviano *et al.* (2009) and consists of two treatment lines, each for 425 L/s. The technology includes: Preliminary treatment + UASB + Maturation pond (two units in series).

Case 3. **Buga city.**

Without reuse. This case study was prepared based on information from different authors (CVC and FAL, 1998; Galvis *et al.*, 2007; Universidad del Valle and CVC, 2007). The technology without reuse (369 L/s) corresponds to: Preliminary + High rate anaerobic lagoon. Buga is located at 969 m.a.s.l. The urban area is 78,059 ha (95%) and the rural area is 4,595 ha (5%). The predominant land use in the Guadalajara river basin is extensive cattle ranching (43%), followed by the cultivation of sugarcane (23%), natural forest (24%), stubble (3%) and other crops (7%).

With reuse. Technology with agricultural reuse was based on Galvis *et al.* (2007) and consists of one treatment line of 369 L/s. The technology includes: Preliminary treatment + UASB + Facultative pond.

5.2.3 Water balance and irrigation requirement

For the three case studies, a water balance between precipitation and evapotranspiration of the sugarcane crop was carried out. With this balance, the amount of water required for irrigation per unit area was obtained (Jaramillo, 2014). Then the effective precipitation was calculated using the method described by Doorenbos and Kassam (1979). For the evapotranspiration of sugarcane cultivation, the methodology of Cruz (2009) was used. The precipitation and evaporation data were obtained from the 'Guachalzambolo', 'El Paraíso' and 'Cenicaña' stations. Precipitation and evaporation (average, monthly and multi-year), were estimated for all cases, based on records corresponding to the period 2002 - 2012.

5.2.4 Available flow for irrigation and cropping area requirements

In each case study, the irrigation area was obtained from the relationship between the flow of waste water produced and the water requirement of the crop per unit area. The location of the irrigation area with treated wastewater was based on knowledge of the study area and five criteria obtained by consulting local experts. These experts were from: irrigation system operators, academics and the environmental authority of the region. These criteria were: 1) Current land use; 2) proximity of sugarcane crops to effluent from wastewater treatment plants; 3) Land slope in direction to the area where the treated wastewater will be used for irrigation; 4) Physical structures that limit the area of irrigation, such as highways, airports and other infrastructure; 5) Vulnerability to contamination of the aquifer system. For the application of this last criterion, the study developed by CVC (1999) on the cartography of intrinsic aquifer vulnerability to pollution using GOD method (Groundwater occurrence, Overall aquifer class and Depth of water table) coupled with a geographic information system (ArcGIS 9.3).

5.2.5 Cost Benefit Analysis (CBA)

Benefits were calculated for the two scenarios (with and without reuse). Common benefits, like health benefits, were not included, but only the incremental costs and incremental benefits were considered. Infrastructure investments were projected for 20 years. The social discount rate (SDR) is the discount rate used in computing the value of funds spent on social projects (Harrison, 2010). The SDR must reflect the 'opportunity cost' that the company attributes to the resources invested in a project in relation to its possible alternative uses. Traditionally, discount rates in Latin American and Caribbean countries are relatively high. National methodologies for obtaining discount rates differ widely. However, in the majority, the exponential discount mechanisms are generally applied through a constant discount rate of 12%, also used by the IDB and other international organizations (Campos *et al.*, 2016). In this study a social discount rate of 11.75% was applied (Comisión de Regulación de Agua Potable y Saneamiento Básico. Ministerio de Vivienda, 2013).

Calculating the costs. The costs were obtained from previous studies on which each case was based. These studies were referenced in the Section 2.2. These costs were projected to reference year 2013, with the Consumer Price Index CPI. EPANET U.S. EPA version 2.0, a modelling software for water distribution system modelling, was used for optimize the sizing of the irrigation distribution networks of sugarcane crops. Information from local institutions and cost models obtained with information about the region (Sánchez, 2013) were used to obtain market prices for the initial investment and the O&M costs of the WWTPs. This same method was used to estimate the costs associated with water supply infrastructure and wells and pumping stations for irrigation of sugarcane crops (Colpozos, 2010). The cost of power consumption was estimated as 0.13 Euros/kW-h. Initial investments include, only for Case 1, the viaduct and the cross-river pumping of the effluent of the Cali wastewater treatment (WWTP-C), to bring treated wastewater from the left bank to the right bank of the Cauca River, to reach sugarcane farms.

Calculating the benefits. The incremental benefits include also avoided cost resulting from the implementation of wastewater reuse. These benefits (avoided costs) have been classified into four groups: 1) Savings from reduced use of fertilizers, based on information specific to sugarcane crops in the Valle del Cauca (Ministerio de Agricultura y Desarrollo Rural, 2010); 2) Reduction in tax for wastewater discharge directly to water bodies, based on regulations of local environmental authority (CVC, 2005); 3) Reduction in payment of the water tariffs (freshwater use) from natural sources (groundwater), as defined by the environmental authority (CVC, 2010); 4) Savings in infrastructure (initial investment) and O&M for irrigation using groundwater. The use of treated wastewater avoids the use of groundwater, thereby avoiding infrastructure and energy costs. To estimate these benefits, the irrigation flow and hydrogeological characteristics of each case were considered. The production of the aquifer system was assessed, which determined the number of wells to be constructed for this water demand. The initial investment costs and O&M of wells for each area were obtained from Colpozos (2010).

Baseline Conditions correspond to year 2013. For Cases 1 and 3 major infrastructure investments were proposed to be made in Year 1. For Case 2, where two WWTPs were considered, there is a first investment in Year 1 (2013) and a second investment in Year 10. The costs and benefits associated with O&M, tax for wastewater discharges (pollution charges) and water tariffs were considered each year from Year 2 (2014) until Year 20 (2033). Costs were obtained in Colombian pesos and a conversion rate of 1 Euro = 2,500 Colombian pesos was used. Based on information specific to sugarcane crops in the Valle del Cauca (Ministerio de Agricultura y Desarrollo Rural, 2010), the following prices for fertilizers were used: NPK = 0.53 Euros/kg and urea = 0.58 Euros/kg. For wastewater discharge tax the following values were used: BOD: 0.0492 Euros/kg and TSS: 0.0211 Euros/kg. For the tariffs for freshwater use, the following values were used: Case 1, WWTP: 0.000823 Euros/m^3; Case 2, Expansion area of Cali: 0.00032 Euros/m^3; Case 3, Buga City: 0.000485 Euros/m^3.

5.2.6 Sensitivity Analysis

Water balance (supply and demand) and irrigation requirement. Nine scenarios were calculated, combining the irrigation requirements obtained for each case study (1.08 L/s-ha, 0.64 L/s-ha y 0.34 L/s-ha).

Water tariffs. 81 scenarios were calculated. Each case study was combined with the three irrigation requirements and with different water tariffs (surface water). These variations were made in a range up to 300 times the value of the rate used in the region, due to the difference between the deviating tariffs applied in Colombia compared to tariffs used in other countries. If we compare the water tariffs of some European countries (England, France, Italy, Spain) with Colombian water tariffs, in terms of the minimum wage, the result is that the tariffs of these countries are more than 300 times the value of the tariff in Colombia.

Tax for wastewater discharge directly to water bodies. 75 scenarios were calculated. Each case study was combined with the three irrigation requirements and different taxes for wastewater discharge directly to water bodies. These variations were performed in a range up to 150 times the baseline tax defined for the study area.

5.3 Results

5.3.1 Water balances and irrigation requirements

The results of the simplified balance (Figure 5.2) were adjusted by the efficiency of the irrigation system, to obtain the actual requirement of the crop. The potential irrigation (period of time) and irrigation requirement estimated for each case study were as follows: Case 1: 334 days/year; 1.08 L/s-ha; Case 2: 62 days/year; 0.34 L/s-ha; Case 3: 212 days/year; 0.64 L/s-ha.

5.3.2 Potential area and irrigation flows

Based on these analyses and the irrigation requirements, the following areas and irrigation flows were defined: Case 1: 3,080 ha and 3,326 L/s; Case 2: 2,530 ha and 880 L/s and Case 3: 576 ha and 369 L/s.

5.3.3 Cost Benefit Analysis (CBA)

The results are presented in *Table 5.2* and *Table 5.3*. These results show the financial viability for Case study 3 (benefits/costs B/C> 1.0) and the financial infeasibility for the case studies 1 and 2 (B/C < 1.0).

5.3.4 Sensitivity analysis

Irrigation requirements (Figure 5.3). For each case, the benefit cost ratio (B/C) was calculated, considering the three irrigation requirements (IR): 1.08 L/s-ha; 0.64 L/s-ha and 0.34 L/s-ha. This includes a relatively broad range of IR and these values are feasible to be presented in the study area. For Case 1, the values of B/C were: 0.60; 0.45 and 0.27, respectively. This showed that for Case 1 the solution with reuse was not feasible for any of these requirements. For Case 2, the values of B/C were: 3.0; 2.0 and 0.8, respectively and for Case 3, the values of B/C were: 2.8; 1.5; 0.8.

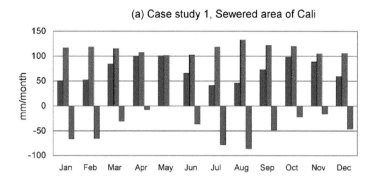

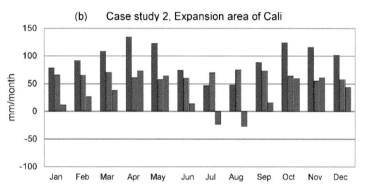

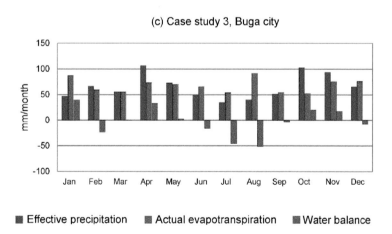

■ Effective precipitation   ■ Actual evapotranspiration   ■ Water balance

Figure 5.2 Simplified water balance of case studies

Table 5.2. Cost and benefits of wastewater management with and without reuse (Euros - 2013)

		Case 1. Sewered area of Cali city (IR: 1.08 L/s-ha)	Without reuse	With reuse	Incremental
Cost	C1	Initial investment-complementary secondary treatment WWTP C.	50,603,871	69,386,789	18,782,918
	C2	Initial investment-transfer system treated wastewater from the left bank to the right bank of the Cauca River	0	2,134,112	2,134,112
	C3	Initial investment - Transport and distribution of water irrigation	0	1,136,898	1,536,898
	C4	Initial investment - Pumping station of treated wastewater	0	409,325	409,325
	C5	O&M - complementary secondary treatment WWTP C.	19,834,932	27,197,173	7,362,241
	C6	O&M - Viaduct to transport related wastewater from the left bank to the right bank of the Cauca River, (Q=3,326 L/s)	0	836,497	836,497
	C7	O&M - network for transport and distribution of water irrigation	0	602,410	602,410
	C8	O&M - pumping station of treated wastewater	0	19,688,201	19,688,201
		Present Value - total costs			*51,332,602*
Benefits	B1	Savings from reduced use of fertilizers	0	3,964,975	3,964,975
	B2	Reduction in tax for wastewater discharge directly to water bodies	0	108,708	108,708
	B3	Reduction in payment of tariff for water use	0	335,262	335,262
	B4	Savings in infrastructure for irrigation using groundwater	0	5,103,300	5,103,300
	B5	Savings in O&M of infrastructure for irrigation with groundwater	0	21,346,013	21,346,013
		Present Value - total benefits			*30,858,258*
		Case 2. Expansion area of Cali city (IR: 0.34 L/s-ha)			
Cost	C1	Initial Investment WWTP - Year 1	3,875,225	4,229,883	354,658
	C2	Initial Investment WWTP - Year 10	1,514,921	1,653,566	138,645
	C3	Initial investment - network for water irrigation - Year 1	-	723,017	723,017
	C4	initial investment - network for water irrigation - Year 10	-	282,645	282,645
	C5	Initial investment - pumping station of treated wastewater Year 1	-	12,041	12,041
	C6	Initial investment - pumping station of treated wastewater Year 10	-	4,707	4,707
	C8	O&M - WWTP	1,965,038	2,144,877	179,839
	C9	O&M - network for transport and distribution of water irrigation	-	283,397	283,397
	C10	O&M - pumping station of treated wastewater benefits	-	262,243	262,243
		Present Value - total costs			*2,241,192*
Benefits	B1	Savings from reduced use of fertilizers	-	362,498	362,498
	B2	Reduction In tax for wastewater discharge directly to water bodies	-	12,384	12,384
	B3	Reduction in payment of tariff for water use	-	5,078	5,078
	B4	Savings in infrastructure for irrigation using groundwater	-	743,811	743,811
	B5	Savings in O&M of infrastructure for irrigation with groundwater	-	710,384	710,384
		Present Value - total benefits			*1,834,155*
		Case 3. Buga city (IR: 0.64 L/s-ha)			
Cost	C1	Initial investment WWTP	3,029,908	4,492,340	1,462,432
	C2	initial investment - transport and distribution of water irrigation	0	276,674	276,674
	C3	Initial investment - pumping station of treated wastewater	0	16,345	16,345
	C4	O&M - WWTP	1,187,617	1,760,839	573,222
	C5	O&M - network for transport and distribution of water irrigation	0	108,446	108,446
	C6	O&M - pumping station of treated wastewater	0	458,925	458,925
		Present Value - total costs			*2,896,044*
Benefits	B1	Savings from reduced use of fertilizers	0	564,387	564,387
	B2	Reduction in tax for wastewater discharge directly to water bodies	0	22,553	22,553
	B3	Reduction in payment of tariff for water use	0	13,299	13,299
	B4	Savings in infrastructure for irrigation using groundwater	0	732,177	732,177
	B5	Savings in O & M of infrastructure for irrigation with groundwater	0	3,001,368	3,001,368
		Present Value - total benefits			*4,333,784*

Social discount rate =11.75%; Time horizon= 20 years; IR: irrigation requirements

Table 5.3 Results of CBA* for the three case studies

Case Study	NPV(Euros)	Benefit/Cost
1. Sewered area of Cali	-20,474,344	0.60
2. Expansion area of Cali	-407,037	0.82
3. Buga city	1,437,740	1.50

*Social discount rate =11.75%; Time horizon= 20 years

Water tariffs and tax for wastewater discharge to water bodies (Figure 5.3). For each case, the values of B/C were calculated, considering the three irrigation requirements IR: 1.08 L/s-ha; 0.64 L/s-ha and 0.34 L/s-ha and modifying the values of water tariffs and tax for wastewater discharges. These variations were performed in a range up to 150 times the baseline (tariff or tax) defined for the study area. The results for the conditions studied are explained below. For Case 1 only financial viability is achieved for IR = 1.08 L/s-ha, but only if the water tariffs increase by approximately 60 times the value of the baseline condition. For cases 2 and 3 there is financial viability for IR = 0.64 L/s-ha and IR = 1.08 L/s-ha. However, for Case 2, when IR = 0.34 L/s-ha this viability is only obtained if tax for water discharges is increased by approximately 40 times the value of the baseline condition or water tariffs are increased by 50 times the value of the baseline condition. For Case 3, when IR = 0.34 L/s-ha viability is only obtained if tax for water discharges is increased by approximately 50 times the value of the baseline condition or water tariffs are increased by 50 times the value of the baseline condition.

5.4 Discussion

The results of the CBA show that in the Upper Cauca river basin there are two key factors that determine the reuse potential of treated wastewater for sugarcane crop irrigation: 1) the water balance and irrigation requirements, and 2) costs corresponding to the management of wastewater for agricultural irrigation, including additional treatment (when it is required) and the infrastructure to bring the treated wastewater to crops. The first of these factors was a key factor for the infeasibility of Case 2 (the expansion area of Cali city), since the requirement for irrigation water was only there for 2 months/year. This time period is very limited to recover the required investments and O&M costs. In Case 1 (the sewered area of Cali city), the irrigation requirements were 11 months/year. For this case the limiting factors for reuse feasibility were investment and O&M costs. These costs included: those associated with wastewater treatment costs and the infrastructure costs required to carry the treated wastewater from the left bank (where the city of Cali is located) to the right bank, where sugarcane crops are to be irrigated. On the other hand, due to the large flow and the specific characteristics of the irrigation area, it was not possible to use the entire effluent of the WWTP. In Case 3, the CBA was positive due to the combination of irrigation requirements (6 months/year) and the supply of treated wastewater. Then the required investments and O&M costs can be recovered by the reuse benefits. Cases 1 and 2 correspond to the same sub-region and between the two potential irrigation areas the distance does not exceed 25 km. However, the irrigation requirements are very different: 1.08 L/s-ha and 0.34 L/s-ha, respectively. Thus, the feasibility of reuse is very dependent on the local conditions of the reuse project (Avellaneda *et al.*, 2004; Moscoso *et al.*, 2002). The specific results obtained in this research cannot be generalized to other contexts.

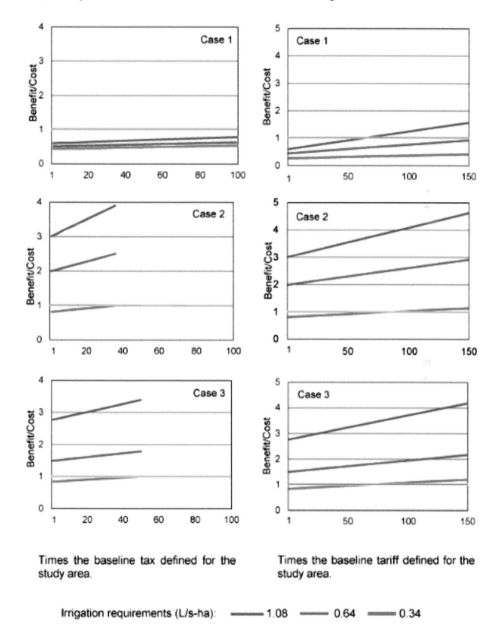

Figure 5.3 Sensitivity Analysis. Effect of tax for water discharges and water tariffs on Benefit /Cost Analysis

The results of sensitivity analysis reinforce the importance of irrigation requirements in the financial viability of reuse. This shows the potential of reuse in areas with water shortage. Different experiences at the global level and documents of international organizations express this relationship as the main driver of the reuse of wastewater in agriculture (Corcoman *et al.*, 2010; Jiménez and Asano, 2008; WHO, 2006; Winpenny *et al.*, 2013). For Case 2, where precipitation limited the reuse of treated wastewater in the expansion area of Cali, the sensitivity analysis showed that with an increase in the irrigation requirements, due to possible changes in rainfall (quantity and temporality) by phenomena such as climate variability or climate change, reuse could become viable in this area.

Sensitivity analysis shows that with the increase of the values the economic instruments (water tariffs and tax for wastewater discharges to water bodies) can favourably affect the CBA results, and with this the reuse feasibility of treated wastewater in sugarcane crops in the Upper Cauca river basin. However, in order to achieve this viability, the value of these economic instruments (for the reference year 2013) they must be increased by many times (Figure 5.3). The fact is however, that compared to western countries, water tariffs and tax for wastewater discharges in Colombia are indeed extremely low, and are likely to increase in the near future. This is because the values used for avoided costs by taxes for water use and taxes for waste water discharges directly to water bodies were negligible, since these unit costs are extremely low. For example, taxes for agricultural irrigation are about 300 times lower, as a percentage of minimum wages, compared with raw water prices in Europe and the United States. Despite this, sugarcane farmers report that irrigation represents between 30% and 60% of the total costs of cultivation (Cruz, 2015).

In the last decade Colombia has been managing its entry into the OECD (Organization for Economic Cooperation and Development (OECD). The OECD countries have seen an increase in their water tariffs for agricultural irrigation use, with the aim of encouraging efficient water use. However, the FAO recognizes that these type of strategies, for other countries, is the result of social, political and economic factors, which makes this kind of action more complex to encourage the reuse of wastewater in agriculture (Winpenny *et al.*, 2013). Other factors to consider in the viability of reuse are associated with the regulations and control of irrigation with treated wastewater. Colombia is among the countries with the highest use of raw wastewater in agricultural irrigation (Jiménez and Asano, 2008), while irrigation with treated wastewater is virtually non-existent (Universidad del Valle and MADS, 2013). Recently the Government of Colombia introduced new regulations for the use of treated wastewater (Ministerio de Ambiente y Desarrollo Sustentable MADS, 2014) through which it aims to encourage reuse in both agricultural irrigation and other types of reuse.

In this research project, the cost-benefit analysis of the options with and without reuse was associated with the irrigation of sugarcane crops. However, other benefits may be included in the integrated wastewater management. These benefits can include the recovery of nutrients and the use of energy (Gijzen and Ikramullah, 1999; Gijzen and Veenstra, 2000). For example, a scheme with: AWWT + lagoons with duckweed + aquaculture + irrigation in agriculture (El-Shafai, 2004), includes additional costs and benefits, which will affect the result of the reuse viability. Other benefits not considered here are the non-market benefits. Wastewater reuse in

agricultural irrigation can contribute to preserving the river's ecological status by reducing the amount of water taken directly from the river for irrigation, while at the same time ensuring a continued water supply for the farmers (Alcon et al., 2010). The wastewater reuse in agriculture increase resources where water scarcity is presented. Its main benefits are related to the economy, the environment and health (Helmer and Hespanhol, 1997). Reuse is a second step of Three-Steps Strategic Approach (Gijzen, 2006; Galvis et al., 2018), where the benefits including pollution control of water bodies, as the direct reduction of water discharge residual water sources (Toze, 2006).

5.5 Conclusions

Feasibility of reuse wastewater in agricultural irrigation can be assessed through the water balance and irrigation requirements. They can be obtained with the following criteria: 1) Current land use; 2) proximity of sugarcane crops to effluent from wastewater treatment plants; 3) Land slope in direction to the area where the treated wastewater will be used for irrigation; 4) Physical structures that limit the area of irrigation, such as highways, airports and other infrastructure; 5) Vulnerability to contamination of the aquifer system. The reuse potential is complemented by the study of financial viability with CBA of the incremental situation when comparing the options with and without reuse of treated wastewater.

The study of the financial viability of the reuse of treated wastewater for the irrigation of sugarcane crops from three case studies in the Upper Cauca river basin in Colombia produced the following results: financial viability (B/C> 1) for Case 3 (Buga city, IR=0.64 L/s-ha) and financial infeasibility (B/C <1) for cases 1 (Sewered area of Cali, IR=1.08 L/s-ha) and 2 (Expansion area of Cali, IR=034 L/s-ha). Case 1 is not feasible even if its IR is increased up to 1.8 L / s-ha, while Case 2 would reach its feasibility condition with a slight increase in its IR. Thus, for an IR = 0.64, its financial viability is guaranteed, with a B/C = 2.0.

The financial viability of reuse in irrigation of sugarcane crops in the upper Cauca river basin is limited because the values of tax for wastewater discharges and water tariffs do not correspond to the values they should have. Thus, for Case 1, only the financial viability for IR = 1.08 L/s-ha is achieved, but only if the water tariff increases approximately 60 times the value of the reference condition. For Case 2, when IR = 0.34 L/s-ha, feasibility is only obtained if the tax for wastewater discharges increases approximately 40 times the value of the baseline condition or if the tariff increases 50 times the value of the baseline condition. For case 3, when IR = 0.34 L/s, financial viability is only obtained if the tax for wastewater discharges increases approximately 50 times the value of the baseline condition or water tariff increases by 50 times the value of the reference condition.

It should be further recognised that Step 2 should be managed and optimised as part of an integrated approach (the 3-SSA), and CBA might give different outcomes if the interventions under the three steps are combined.

5.6 References

Alcon, F., Pedrero, F., Martin-Ortega, J., Arcas, N., Alarcón, J., and De Miguel, M. (2010). The non-market value of reclaimed wastewater for use in agriculture: a contingent valuation approach. *Spanish Journal of Agricultural Research* 8(S2), pp 187-196.

Amigos de la Tierra América Latina y el Caribe (2016). Estado del Agua en América Latina y el Caribe, 1 Edición. ed.

Asano, T., Burton, F., Leverenz, H., Tsuchihashi, R. and Tchobanoglous, G. (2007). Water reuse: Issues, technologies and applications. AECOM Press & McGraw Hill Professional, United States.

Avellaneda, M., Bermejillo, A. and Mastrantonio, L. (2004). Aguas de riego: calidad y evaluación de su factibilidad de uso. EDINUC, Mendoza, Argentina.

Ayers, R. and Wescot, D. (1987). La calidad del agua en la agricultura. Estudio Riego y Drenaje No. 29. FAO, Rome, Italy.

Becerra, C., Lopes, A., Vaz, I., Silva, E., Manaia, C. and Nunes, O. (2015). Wastewater reuse in irrigation: A microbiological perspective on implications in soil fertility and human and environmental health. *Environ. Int.* 75, 117–135.
https://doi.org/https://doi.org/10.1016/j.envint.2014.11.001.

Bixio, D. and Wintgens, T. (2006). *Water Reuse System Management Manual AQUAREC*. Office for Official Publications of the European Communities, Luxemburg.

Boardman, A., Greenberg, D., Vining, R., Weimer, D. (2001). Cost-benefit Analysis: Concepts and Practice., 2nd Editio. ed. Prentice Hall, San Diego, USA.

Brega Filho, D. and Mancuso, P. (2003). Conceito de reúso de água. En: Reúso de Água, Universidade de São Paulo - Facultade de Saúde Pública. ABES, Brasil, 579.

Campos, J., Serebrisky T. and Suárez-Alemán A. (2016). Tasa de descuento social y evaluación de proyectos: algunas reflexiones prácticas para América Latina y el Caribe. Banco Interamericano de Desarrollo.

Candela, L., Fabregat, S., Josa, A., Suriol, J., Vigues, N. and Mas, J. (2007). Assessment of soil and groundwater impacts by treated urban wastewater reuse. A case study: Application in a golf course (Girona, Spain). Sci. *Total Environ.* 374, 26–35.
https://doi.org/https://doi.org/10.1016/j.scitotenv.2006.12.028.

Capra, A. and Scicolone, B. (2007). Recycling of Poor Quality Urban Wastewater by Drip Irrigation Systems. *Clean. Prod.* 1529–1534.
https://doi.org/https://doi.org/10.1016/j.jclepro.2006.07.032.

Colpozos (2010). Personal communication about the cost model of pumping systems and irrigation in Cauca valley.

Comisión de Regulación de Agua Potable y Saneamiento Básico. Ministerio de Vivienda, (2013). Resolución CRA 628 de 2013. Definición de la tasa de descuento aplicable a los servicios públicos domiciliarios de acueducto y alcantarillado.

Corcoman, E., Nellemann, E., Bos, R., Osborn, H. and Savelli, H. (2010). Sick water? The central role of wastewater management in sustainable development, First Edit. ed. Earthprint, Norway, Nairobi-Kenya.

Cruz, J.R. (2009). Medición del agua de riego. Tecnicaña, 34, 27-24.

Cruz, J.R. (2015). Manejo eficiente del riego en el cultivo de la caña de azúcar en el valle geográfico del río Cauca. Centro de Investigaciones de la Caña de Azúcar. Cenicaña, Santiago de Cali, Colombia.

CVC (1999). Evaluación de la vulnerabilidad a la contaminación de las aguas subterráneas en Valle del Cauca. Grupo de Recursos hídricos CVC, Cali, Colombia.

CVC (2005). Acuerdo 018 de 2005, por el cual se establece la tarifa de cobro por concepto de tasa retributiva.

CVC (2010). Bases de datos de usuarios del recurso hídrico en el Valle del Cauca. Cali, Colombia.

CVC and FAL (1998). Sistema de Información Ambiental, cartografía básica. Cali, Colombia.

Doorenbos, J. and Kassam, A. (1979). Yield response to water, FAO Irrigation and Drainage Paper No. 33.

El-Shafai, S. (2004). Nutrient valorization via duckweed based wastewater treatment and Aquaculture. UNESCO-IHE Institute-Wageningen University.

EMCALI (2007)a. Plan de Saneamiento y Manejo de Vertimientos 2007 - 2016. Unidad Estrategica de Negocios de Acueducto y Alcantarillado. Candelaria-Valle del Cauca.

EMCALI (2007)b. Resultados de investigación en agua potable, aguas residuales y biosólidos. Cali, Colombia.

EMCALI (2011). Informe de vertimientos realizados por el municipio de Cali. Periodo: 2000 - 2010. In: Seminario acerca de la nueva normativa de vertimientos. Decreto 3930 de 2011, Universidad ICESI. Cali, Colombia.

EMCALI and Hidroccidente (2006). Estudio de Alternativas de Dotación de Los Servicios Públicos de Acueducto, Alcantarillado y Complementario de Alcantarillado En La Zona de Expansión de La Ciudad de Cali Denominada 'Corredor Cali-Jamundí'. Cali, Colombia.

EMCALI and Consorcio Ingesam-Hidroccidente (2009). Estudios y diseños de las redes de acueducto y alcantarillado de la primera etapa de expansión del corredor Cali-Jamundí. Cali, Colombia.

Galvis, A., Cardona, D. and Aponte, A. (2007). Technology selection for pollution control and wastewater impact reduction in Buga, Colombia, in: 2nd SWITCH Scientific Meeting. SWITCH, Tel-Aviv, Israel.

Galvis, A., Zambrano, D., P., Van der Steen., P. and Gijzen, H. (2014). Evaluation of a pollution prevention approach in the municipal water cycle. *Clean. Prod.* 66, 599–609. https://doi.org/https://doi.org/10.1016/j.jclepro.2013.10.057.

Galvis, A., Van der Steen, P. and Gijzen, H. (2018). Validation of the Three-Step Strategic Approach for Improving Urban Water Management and Water Resource Quality Improvement. *Water, MDPI Journal*, 10, 188; doi:10.3390/w10020188,

Gaviano, A., Zambrano, D., Galvis, A. and Rousseau, D. (2009). Application of natural treatment systems for wastewater pollution control in the expansion area of Cali, Colombia., in: AGUA. Cinara-Universidad del Valle, Cali, Colombia.

Gijzen, H. (2001). Anaerobes, aerobes and phototrophs (A winning team for wastewater management). *Water Sci. Technol.* 44, 123–132.

Gijzen, H. (2006). The role of natural systems in urban water management in the City of the Future - A 3-Step Strategic Approach. *Ecohydrol. Hydrobiol.* 6, 115–122. https://doi.org/https://doi.org/10.1016/S1642-3593(06)70133-1.

Gijzen, H. and Ikramullah, M. (1999). Pre-feasibility of duckweed-based wastewater treatment and resource recovery in Bangladesh. Washington, USA.

Gijzen, H. and Veenstra, S. (2000). Duckweed Based Natural Systems for Wastewater Treatment and Resource Recovery, in: Olguin, E., Sanchez, G., Hernández, E. (Eds.), Environmental Biotechnology and Clean Bioprocesses. CRC Press, Veracruz, México, pp. 83–100.

Harrison, M. (2010). Valuing the Future: the social discount rate in cost-benefit analysis.

Helmer, R., and Hespanhol, I. 1997. Water pollution control. A guide to the use of water quality management priciples, WHO and UNEP Ed., UK.

Hespanhol, I. (2003). Potencial de reúso de água no Brasil, in: Mancuso, P., Santos, H. (Eds.), Reúso de Água. Universidade de Sáo Paulo Faculdade de Saúde Pública, Sao Paulo-Brazil.

Jaramillo, M.F. (2014) Potencial de Reuso de Aguas Residuales Domesticas como Estrategia de Prevención y Control de la Contaminación en el Valle Geográfico del rio Cauca. MSc Thesis, Universidad del Valle, Cali, Colombia.

Jimenez, B. and Asano, T. (2004). Acknowledge all approaches: The global outlook on reuse. *Water 21. Mag. Int. Water Assoc.* 32–37.

Jiménez, B. and Asano, T. (2008). Water Reuse: An International Survey of Current Practice, Issues and Needs, International WATER ASSN. IWA Publishing.

Jimenez, B., Mara, D., Carr, R. and Brissaud, F. (2010). No Title, in: Dreschsel, P., Scott, C., Raschid-Sally, L., Redwood, M., Bahri, A. (Eds.), Wastewater Irrigation and Health: Assessing and Mitigating Risk in Low-Income Countries. Earthscan, IDRC and IWMI.

Ministerio de Agricultura y Desarrollo Rural (2010). Evaluación Departamental de Costos de Producción de Caña de Azucar.

Ministerio de Agricultura y Desarrollo Rural (2011). Desarrollo de capacidades en el uso seguro de las aguas residuales para la agricultura. Colombia.

Ministerio de Ambiente y Desarrollo Sustentable MADS (2014). Resolución N° 1207 July 25, 2014. Por la cual se adoptan disposiciones relacionadas con el uso de aguas residuales tratadas.

Moscoso, J., Egocheaga, L., Ugaz, R. and Trellez, E. (2002). Sistemas integrados de tratamiento y uso de aguas residuales en América Latina: realidad y potencial. Lima, Perú.

Peña, M. (2002). Advanced Primary Treatment of Domestic Wastewater in Tropical Countries: Development of High Rate Anaerobic Ponds. University of Leeds.

Pescod, M. (1992). Wastewater treatment and use in agriculture - FAO irrigation and drainage paper 47. Newcastle, UK.

Ramirez, C., Santacruz, S., Bocanegra, R. and Sandoval, M.C. (2010). Incidencia del embalse de Salvajina sobre el régimen de caudales del río Cauca en su valle alto. *Ing. los Recur. Nat. y del Ambient.* 9, 89–99.

Sánchez, F. (2013). Detailed costs for some WWTPs designed and built by Sánchez Engineering.

Sandoval, M. and Ramírez, C. (2007). El río Cauca en su Valle alto. Un aporte al conocimiento de uno de los ríos más importantes de Colombia. UNIVALLE-CVC, Santiago de Cali, Colombia.

Sandoval, M., Ramirez, C.A., and Santacruz, S. (2007). Optimización de la regla mensual de operación del embalse de Salvajina. *Ing. los Recur. Nat. y del Ambient.* 6, 93–104.

Toze, S. (2006). Reuse of effluent water. Benefits and risks. *Agriculture Water Management*, 80(1-3), 147-159.

UNESCO - IHP (2014). Water in the post-2015 development agenda and sustainable development goals.

Universidad del Valle and CVC (2007). Selección de tecnología y predimensionamiento hidráulico del sistema de tratamiento para las aguas residuales de Buga, Colombia.

Universidad del Valle and EMCALI (2008). Estudio de evaluación de la potencialidad del reuso para la agricultura del efluente de la PTAR-C. Fase 2. Eidenar. Cali, Colombia.

Universidad del Valle and MADS (2013). Evaluación del estado de los procesos de reuso de aguas residuales en Colombia.

U.S. EPA (2012). Guidelines for Water Reuse.

WHO (2006). Guidelines for the safe use of wastewater, excreta and grey water. Volume 2. Wastewater use in agriculture.

Winpenny, J., Heinz, I. and Koo-Oshima, S. (2013). Reutilización del agua en la agricultura. Beneficio para todos. Informe sobre temas hídricos, First Edit. ed. FAO, Roma, Italia.

Chapter 6

Comparing dynamic and steady-state modelling to study the impact of pollution on the Cauca River

Source: CVC photo file

This chapter is based on:
Galvis, A., Hurtado, I, Martinez-Cano, C., Urrego, J.G., Van der Steen, N.P. and Gijzen, H.J.
Dynamic condition approach to study the self-purification capacity of water bodies: the case of
the Cauca river and Salvajina dam, Colombia. Presented in 11th International Conference on
Hydroinformatics HIC 2014, New York City, USA, August 17-20, 2014.

Abstract

The three Steps Strategic Approach (3-SSA) consisting of: 1) minimization and prevention, 2) treatment for reuse and 3) stimulated natural self-purification has been proposed as an integrated approach to efficiently address the problems associated with wastewater discharges and its impact on receiving water bodies. This study focuses on Step 3, comparing dynamic and steady state conditions of quantity and quality (DO and BOD_5) in the Cauca River, Colombia, and the impact of pollution and wastewater discharges. Over the past 60 years, the Cauca River has lost much of its natural self-purification capacity, represented in the wetlands of its floodplain. A multipurpose reservoir (Salvajina dam) was built in 1985 for pollution control (dilution capacity), power generation and flood control. The results of this study show that self-purification capacity in the Cauca River is strongly affected by abrupt changes in hydraulic flows, especially due to the operation of Salvajina reservoir and the type and size of the received pollution from point-source and non-point source pollutants. The results also show the importance of considering the dynamic behaviour of the system under Step 3 of the 3-SSA and the importance of improving policies and regulations for monitoring and pollution control.

6.1 Introduction

In order to be able to efficiently address problems caused by municipal wastewater discharge, it is important to adopt an integrated approach that includes control of contamination at source, followed by treatment and responsible discharge or reuse of final effluent (Abbott *et al.*, 2013). These 'cleaner production' principles have been successfully applied in the industrial sector and now these concepts are being applied to integrated water resources management. In this context, the conceptual model of the Three Step Strategic Approach (3-SSA) was developed, consisting of: 1) pollution prevention and waste minimization 2) treatment for reuse and 3) stimulated natural self-purification (Gijzen, 2006). When a river is polluted, the water quality deteriorates, limiting water use and ecosystem functions (González *et al.*, 2012). However, the self-purification capacity of a river allows it to restore (partially or fully) its quality through re-aeration and natural processes of biodegradation (Von Sperling, 2005). The mechanisms of self-purification can take the form of dilution of polluted water with an influx of surface or groundwater or through a combination of complex hydrological, biological and chemical processes (Vagnetti *et al.*, 2003; Ifabiyi, 2008; Ostroumov, 2008).

Although there are multiple contaminants (point and diffuse pollution) that may exist in the aquatic environment, the availability of water quality data for dynamic analysis is generally limited. In this study dissolved oxygen (DO) and biochemical oxygen demand (BOD_5) are presented as classic indicators of water pollution. The DO concentration is a primary measure of a stream's health, and it responds to the BOD_5 load (Khan and Singh, 2013). This is why oxygen demand (OD) has been traditionally used to assess the pollution degree and the oxygen profile downstream from a point source of pollution indicates the self-purification capacity of water bodies. Its measurement is simple; however, the complex mechanisms involved in DO must be studied by mathematical modelling (Von Sperling, 2005). Streeter and Phelps (1925) developed the first models of balance between the DO supply rate from re-aeration and the DO consumption rate from stabilization of organic waste. This deoxygenation rate has been expressed as an empirical first order reaction, producing the classic DO sag model. The sag model is usually a steady-state model. However, temporal variability defines two categories of models: the steady-state, where the variables describing the system are considered constant over time, and dynamic models, using fluctuating values. Selection of a model will depend on the study objectives, specific characteristics of the study site and the availability of information (IDEAM, 2011).

In Colombia and other Latin American countries, investments to recover rivers and improve the water quality for their various uses have focused on 'end of pipe' approaches via the construction of waste water treatment plants (WWTP). The water pollution control usually only considers point sources of pollution and steady-state conditions of quantity and quality in the pollution sources and the receiving water bodies. However, both the rivers and their tributaries, especially those associated with the urban drainage system, can present very dynamic behaviour. In rivers with reservoirs for flow regulation, there may be flow variations of over one hundred percent with in a daily cycle. On the other hand, the diffuse pollution associated with urban runoff (first flush effect) and the re-suspension of settled material in the urban drainage system generates pollution peaks. In these cases, in a few hours, the pollution load received by the river may be higher than load discharged by point pollution throughout a whole day. This is the case in the Upper Cauca River basin in Colombia. In this basin, where approximately half of the municipalities already have a WWTP, water quality is now worse, compared to the time when these treatment systems did not exist. This deterioration can be explained by the pollution load increase (domestic and industrial) generated in the basin and the limited effectiveness of wastewater treatment plants, including the one in Cali City (WWTP-C). However, this deterioration is also associated with variations in the Cauca River flow due to the upstream Salvajina reservoir and the impact of its tributaries, in particular the South Drainage System of the City of Cali. To analyse these impacts, a study was carried out using dynamic modelling.

Effective pollution control requires an integrated approach such as the 3-SSA (Gijzen, 2006), including minimization and prevention of waste (Step 1), treatment for reuse (Step 2) and maximising self-purification capacity of the receiving water body (step 3). This study focuses on Step 3 and aims to contribute to improve the policies of monitoring and pollution control of water resources and the criteria to prioritize infrastructure investments for the effective control of water pollution, including the construction of WWPTs.

6.2 Methodology

6.2.1 Study area

The Cauca River is the second most important fluvial artery of Colombia and the main water source of the Colombian southwest. It has a length of 1,204 km with a basin of 59,074 km². The study area is the Upper Cauca river basin (Figure 6.1). This stretch of the river has an average width of 105 m. The depth can vary between 3.5 and 8.0 m. The longitudinal profile of the river shows a concave shape and a hydraulic slope, which varies between 1.5×10^{-4} m/m and 7×10^{-4} m/m (Ramirez et al., 2010). The sugar cane crops and the Colombian sugar industry are located in the flat area along the Upper Cauca river basin. In the mountain area, there are coffee crops and associated industry. There are also other farming developments, and other economic activities such as mining and manufacturing.

The Cauca River has been used for fishing, recreation, energy generation, riverbed matter extraction, human consumption, irrigation and industry. The Salvajina reservoir started operations in 1985 and is part of the flow regulation project of the Cauca River, implemented for flood control, improvement of water quality and power generation (Galvis, 1988). The reservoir power station has a capacity of 270 MW. The reservoir operates between levels of 1,110 and 1,150 meters above sea level (m.a.s.l.), it has a minimum discharge of 60 m³/s and an average daily flow rate of 140 m³/s in the Juanchito Station (Sandoval *et al.*, 2007). The Cauca River is also used as a receiving water body for solid waste and dumping of industrial and domestic wastewater, which contributes to the deterioration in water quality. In the study area, there are currently 3.8 million inhabitants who form the source for approximately 134 T/d of BOD₅ to the Cauca River in the study reach. In addition to organic matter (measured in terms of BOD₅), the river has other types of associated contaminants with acute risk (coliforms and turbidity) and chronic risk (colour, phenols, heavy metals, pesticides and emerging pollutants).

6.2.2 Data collection for steady-state and dynamic conditions

The Cauca river has 15 monitoring stations in the 'La Balsa -Anacaro' reach. There is more than 50 years of quantity (flows, levels, cross sections) and quality (temperature, DO, BOD₅) data available. This information can be classified into two major periods: before and after the Salvajina dam construction. In addition, we used information available on the water quality and quantity in the main tributaries of the Cauca River, such as wastewater discharges and water intake for domestic and industrial uses. Although most of data correspond to steady-state condition, there is also information available to represent the dynamic condition of the Cauca River and its main tributaries in the study area.

Reference year and base line condition.
The year 2014 was taken as reference year, with a mean discharge of 143 m³/s at Juanchito Station and with a level in the Salvajina reservoir at 1,145 m.a.s.l. In 2014 Cali and some other municipalities already had wastewater treatment plants in operation. As for the Salvajina reservoir operation, information on water quality, levels, effluent flows and power generated was used (Hurtado, 2014). The base line for steady-state and for dynamic conditions correspond to the dry season of the reference year. The Table 6.1 presents BOD₅ discharged to Cauca River for the base line along the Hormiguero - Anacaro reach.

Table 6.1 BOD₅ discharged to Cauca River. Base line, dry season, 2014, steady-state condition

Reach in Cauca River	Tributaries	BOD₅ (T/day)	Total for the reach (T/day, % of total)
La Balsa - Hormiguero	Palo river	4.48	
	Jamundí river	1.00	
	Other	2.37	7.9 (5.8%)
Hormiguero - Mediacanoa	South Channel (Cali city)	1.61	
	WWTP-C (Cali city)	51.80	
	Industrial zone	13.27	
	Yumbo river	2.22	
	Guachal river	8.15	
	Cerrito river	7.25	
	Other	20.97	105.3(78.5%)
Mediacanoa - Anacaro	Tulua river	7.74	
	Morales river	1.20	
	Other	12.05	21.0 (15.7%)
Total		**134.11**	**(100%)**

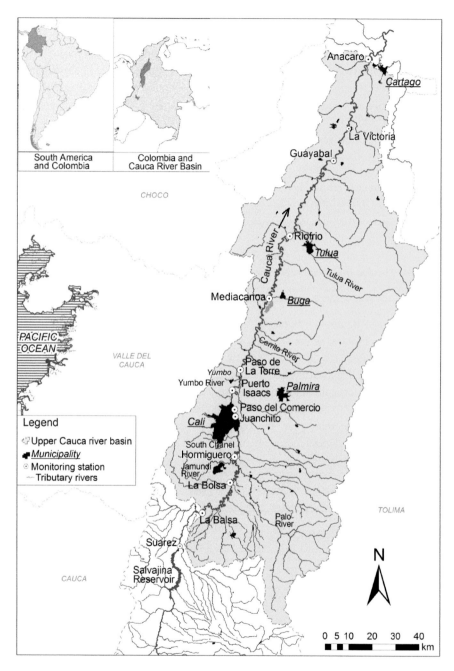

Figure 6.1 Upper Cauca river basin, Salvajina - Anacaro stretch
Adapted from Universidad del Valle and CVC (2009)

6.2.3 Implementation of the model

The hydrodynamic and water quality model of the Cauca River was implemented in the MIKE 11 model. The Cauca River has 15 monitoring stations in the La Balsa–Anacaro stretch (Figure 6.1). The calibration and validation of the quantity (roughness) and quality (BOD, DO) components to apply the MIKE 11 model were based on (Universidad del Valle and CVC, 2007). The model consists of 387 cross sections, 2 external boundaries: La Balsa (km 27.4) and Anacaro (km 416.1), 95 internal boundaries which include 34 rivers and streams, municipal wastewater discharges, 24 industrial wastewater discharges and 36 water extraction sites. Two monitoring campaigns were used: calibration (2005) and validation (2003). The quality component of the MIKE 11 model at Level 1 and the Churchill equation for the re-aeration calculation were selected. Then the values resulting from the calibration-validation process are presented: Strickler roughness ($m^{1/3}$ s^{-1}); BOD_5 degradation constant (d^{-1}) and Benthic Oxygen Demand (g $O_2/m^2/d$). The values are presented in this order for each monitoring station on the Cauca River, the La Balsa–Anacaro stretch: La Balsa (40; 0.15; 1.5); La Bolsa (20; 0.15; 2) Hormiguero (40; 0.3; 3); Juanchito (33; 0.4; 5); Puerto Isaacs (60; 0.35; 5); Paso de la Torre (60; 0.33; 3); Mediacanoa (34; 0.2; 2); Guayabal (30; 0.17; 1); La Victoria (33; 0.17; 1) and Anacaro (32; 0.17; 1).

The model was applied for steady-state or dynamic condition, depending on the type of scenario to be studied. The simulation was set at 10 days, considering the period required for the stabilization of the model and the average time of water travel between La Balsa and Anacaro stations in the Cauca River. Dynamic effects studied in this paper were calculated and displayed for a period of three days, because they were associated with dynamic activities having a cycle of daily changes. This includes the following situations: 1) typical summer day (dry season) variations in reservoir cycling (Salvajina reservoir) during 24 hours; 2) waste water discharges variations of Cali city (WWTP-C) also have a daily variation; 3) peaks of pollution runoff surface from South Drainage System (SDS) have a duration of, usually, less than a day.

6.2.4 Formulation of scenarios

The reservoir operation between 1985 and 2014 was analysed. Typical behaviours of the reservoir were studied in relation to operational policies and criteria set by the power generation company and the environmental authority. High, medium and low water levels of the reservoir were defined and each of these levels were associated with typical series of power generation and outflow of the reservoir. In turn these time series were associated with discharge time series in the upstream boundary (La Balsa) and in the reference station (Juanchito). An analysis of available data demonstrated typical combinations of water levels in the reservoir with discharges in the Cauca River for typical winter (wet season), summer (dry season) and transition conditions (Universidad del Valle & CVC, 2009, Ramirez et al., 2010; CVC, 2013; EMCALI, 2013; EPSA, 2014). With this analysis, the number of studied scenarios was reduced. The base line (steady-state and dynamic conditions) and hypothetical scenarios were formulated in order to study the possible behaviour of the river with and without the effect of the Salvajina reservoir: steady-state condition without effect of Salvajina (S1); dynamic conditions: WWTP-C is out of operation (S2) and impact of rainfall event in the South Drainage System of Cali city (S3 and S4).

Steady-state conditions.
The base line was formulated to compare the impact of the Salvajina reservoir effect on the water quality of the Cauca river, with the assumption of permanent flow (95% in the flow duration curve at the Juanchito Station) and steady-state condition in both the upstream borders (La Balsa Station) and the tributaries (rivers and wastewater discharges). After construction of the Salvajina reservoir (Base line, Post-Salvajina condition) the flow at Juanchito Station corresponds to 143 m^3/s approximately. Before the construction of the Salvajina reservoir (Scenario S1, pre-Salvajina condition) the flow at Juanchito Station corresponded to 88 m^3/s.

Dynamic conditions.
For these conditions, the following cases were modelled: dynamic base line, S2: WWTP-C out of operation, S3: pollution peak coincides with the minimum flow in the Cauca river and S4: pollution peak coincides with the maximum flow rate in the Cauca river. This study analysed data obtained through measurements under dynamic conditions carried out over the last decade. Figure 6.2 shows the typical load variation in the WWTP-C and the Figure 6.3 presents the BOD$_5$ load produced by a rainfall event (first flush effect; August 22-23, 2003) and measured in the South Drainage System of Cali (South Channel). Such events have been frequently observed and force the municipality to temporarily close river water intake to the water supply system for two million inhabitants in Cali (Moreno, 2014).

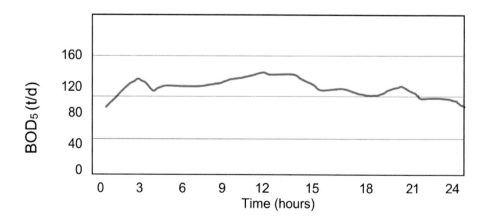

Figure 6.2 BOD$_5$ influent load to WWTP-C. Baseline, dry season 2014. Flow Juanchito St.= 143 m^3/s
Source: EMCALI (2013)

These closures have a duration from a few hours up to two days. Pollution peaks coming from the South Channel discharge into the Cauca River are associated with diffuse pollution sources such as runoff from rural and urban areas. Additionally, the re-suspension of sediments and solid waste accumulated in the drainage network also contributes to the occurrence of pollution peaks. It was assumed that dynamic flow condition at the upstream boundary of the Cauca River (La Balsa Station) only depends on the Salvajina reservoir (Figure 6.4).

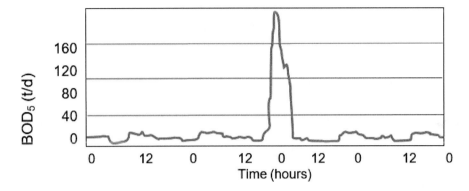

Figure 6.3 BOD$_5$ load discharged into the Cauca river during a rainfall event in the South Channel drainage area, August 22-23, 2003
Source: Universidad del Valle and CVC (2004)

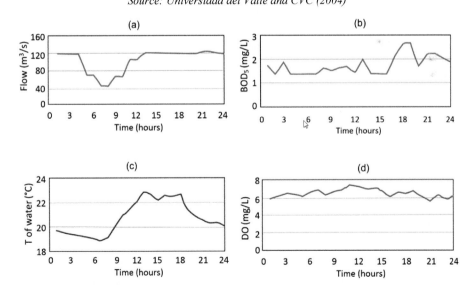

Figure 6.4 Cauca River under dynamic conditions at La Balsa Station. Dry season, average flow at Juanchito Station 143 m³/s

Based on (a): Universidad del Valle and CVC (2009) and EPSA (2014); (b, c and d): Universidad del Valle and CVC (2004, 2009).

6.3 Results and discussion

Base line steady-state condition and Scenario 1. Impact of Salvajina reservoir considering steady-state condition

The modelling results with MIKE 11 for steady-state conditions (quantity and quality) were: 1) Base line for steady-state condition (flow in Juanchito Station: 143 m³/s) corresponds to post-Salvajina condition (after construction of the dam, in 1985) minimum DO (0.5 mg/L) was found at Puerto Isaac Station (km 155); 2) For S1 corresponds to pre-Salvajina condition (before construction of the dam, 1985) corresponding to flow in Juanchito Station: 88 m³/s and minimum DO= 0.3 mg/L was found at this same station.

Base line dynamic condition and Scenario 2
Impact of WWTP-C out of operation considering a dynamic condition. This scenario is characterized by mean flow of 143 m³/s at Juanchito station (km 139) and considering flow variation, water temperature, BOD₅ and DO in the upper boundary condition (La Balsa Station). The Figure 6.5Figure 6.5 The DO in the Cauca River at three stations for base line dynamic conditions and with WWTP-C in operation and average flow of Juanchito St.= 143 m3/s (Base line Dynamic condition) shows the results for modelling of the base line, with WWTP-C operating and Figure 6.6Figure 6.6 The DO in the Cauca River at two stations for dynamic conditions when the WWTP-C was out of operation and average flow of Juanchito St.= 143 m3/s (Scenario S2) shows the results for modelling of S2, considering WWTP-C is out of operation, thus its inflow discharges directly to the Cauca River. This condition implies that in S2 the pollutant load discharged to the Cauca River is increased by 37 T/day (Figure 6.6) In the Figure 6.5 the anaerobic condition of the Cauca River is presented at the station of Port Isaac (km 155), while in second scenario (Figure 6.6) the anaerobic condition is obtained at the Paso del Comercio station (km 145).

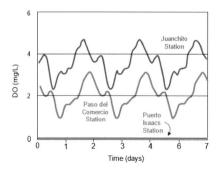
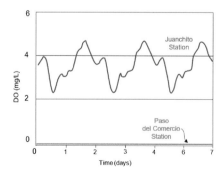

Figure 6.5 The DO in the Cauca River at three stations for base line dynamic conditions and with WWTP-C in operation and average flow of Juanchito St.= 143 m³/s (Base line Dynamic condition)

Figure 6.6 The DO in the Cauca River at two stations for dynamic conditions when the WWTP-C was out of operation and average flow of Juanchito St.= 143 m³/s (Scenario S2)

Scenarios S3 and S4 in dynamic condition
The results of a rainfall event in the South Drainage System of Cali city are shown in the Figure 6.7. For S3 the pollution peak (see Figure 6.3) coincides with the lowest flow of the Cauca River at the point of discharge. Scenario S4 is defined by a pollution peak that coincides with the highest flow of the Cauca River at the point of discharge.

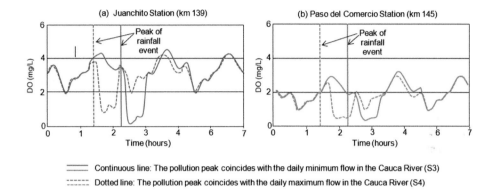

Continuous line: The pollution peak coincides with the daily minimum flow in the Cauca River (S3)
Dotted line: The pollution peak coincides with the daily maximum flow in the Cauca River (S4)

Figure 6.7 The DO in the Cauca River at two stations for dynamic conditions showing the impact of a rainfall event in the South Drainage System of Cali city for average flow of Juanchito St.= 143 m³/s (Scenarios S3 y S4)

The Table 6.2 summarizes the results of modelling of the different scenarios. For steady-state flow condition, for the base line and S1 (Pre-Salvajina dam construction). Under this flow condition, in theory, a positive effect on self-purification capacity is produced by increasing the dilution capacity. DO at the reference Station Juanchito (point of intake of the water supply system of Cali city) increases from 3.0 mg/L (S1) to 4.1 mg/L (base line) and the minimum DO (Puerto Isaacs Station) increases from 0.3 mg/L (S1) to 0.5 mg/L (base line). In the past, this type of analysis proved the benefits of using the Salvajina reservoir for water quality improvement in the Cauca River (Galvis, 1988). However, this condition (steady-state flow) does not match with the measurements in the Cauca River after the start-up of the Salvajina dam, in 1985. The model makes it possible to increase the average flow in Juanchito Station, for example from 88 m³/s (S1) to 143 m³/s (base line), with high flows during the day (300 m³/s), but maintaining (or decreasing) the minimum (70 m³/s or less) at night. Measurements show an increasing impact of the dynamic conditions of pollution discharges on the water quality of the Cauca River (Universidad del Valle and EMCALI, 2006; Universidad del Valle and CVC, 2007; Velez *et al.*, 2014). For example, the pollution peaks coming from the South Drainage System affect the water quality in the Cauca River and force the closure of the water supply system of Cali city. These closures were increased from 10 in 2000 to 41 in 2013 (Moreno, 2014).

Table 6.2 The DO at the Juanchito Station and minimum DO at the Cauca River. Summary of the results of baseline and scenarios modelling for an average flow of Juanchito Station= 143 m³/s

Scenario			Condition	Average Flow Q Juanchito [1] (m³/s)	DO Juanchito[3] Station (mg/L)	DO min (mg/L) Station (Abscissa) [1]
Impact of Salvajina reservoir	S1	Steady Conditions	Base Line (steady-state) Post-Salvajina dam construction	143	4.1	0.5 Puerto Isaacs (km 155)
			Pre-Salvajina dam construction	88	3.0	0.3 Puerto Isaacs (km 155)
Impact of WWTP of Cali city	S2	Dynamic Conditions	Base Line (Dynamic)	143	2.1-4.6	0.0 Puerto Isaacs (km 155)
			WWTP [4] of Cali is out of operation	143	2.1-4.6	0.0 P. del Comercio (km 145)
Impact of Rainfall event in the South Channel drainage	S4		Pollution peak SDS[2] coincides with the daily minimum flow in the Cauca River	143	0.0-4.2	0.0 Juanchito (km 139)
	S4		Pollution peak coincides with the daily maximum flow in the Cauca River	143	0.8-4.2	0.0 P. del Comercio (km 145)

(1) Outlet of Salvajina reservoir: km 0.0
(2) Discharge of SDS (South System Drainage): km 127.7
(3) Juanchito Station (km 139.3); Intake of the water supply system of Cali is located at km 139.0
(4) Discharge of WWTP-C: km 142.1

For dynamic conditions, the water quality modelling corresponding to the base line 2014 shows the DO variations in Juanchito Station between 2.1 and 4.6 mg/L, but in Puerto Isaacs Station DO is almost zero. The WWTP-C removes approximately 37.2 t/d of BOD_5. However, this pollutant load reduction was not sufficient to prevent anaerobic conditions at the critical point (Puerto Isaacs Station), where a mean flow below 143 m3/s in a dynamic condition is generated by the Salvajina reservoir operation. Additionally, when the WWTP-C was out of operation (S2), the modelling results showed a critical DO between 2.1 and 4.6 mg/L at Juanchito (station before WWTP-C effluent discharge). For scenario S2, DO is zero in Paso del Comercio Station, located 10 km upstream from Puerto Isaacs Station.

Due to the runoff dynamic effects and re-suspension of sediments in the urban drainage network, the modelling results vary when the pollution peak of the discharge coincides with low flows (S3) or if this peak coincides with periods of Cauca River high flows (S4). The results for S3 and S4 were more critical than those for S2 (WWTP-C out of operation), since they produce a DO value that is practically zero at Juanchito and Paso del Comercio Stations. Additionally, S3 and S4 DO ranges at Juanchito were more critical than S2 DO ranges. The modelling results were consistent with low DO levels measured at Juanchito Station, closure of the water supply system of Cali and anoxic conditions of the Cauca River downstream of this station (Moreno, 2014). These results show the limited impact of 'end of pipe solutions' that consider WWTPs as the only strategy for improving the water resources quality.

Self-purification is the recovery process of water resources after an organic pollution episode. In this process, organic compounds are diluted and progressively transformed by microbial and biochemical decomposition. In rivers, the self-purification capacity depends mainly on: 1) the flow, which dilutes the pollution discharge and facilitates its subsequent degradation to reduce its toxicity, 2) water turbulence, which provides oxygen to the water favouring microbial activity, and 3) the nature and size of the discharges (Vagnetti, 2003; Von Sperling, 2005). The self-purification capacity will also depend on DO levels. Once a water body turns anoxic the self purification capacity is decimated, Other factors include the presence and activity levels of algae, which may enhances the microbial decomposition processes during day-time due to higher DO levels. The self-purification capacity of a water body can further be influenced via ecohydrology interventions, which presents additional strategic options for cost-effective water quality improvement, as proposed under step 3 of the 3-SSA.

The self-purification capacity of the Cauca River is strongly affected by abrupt changes in its dilution ability and the type and size of the received pollution from point source and non-point source pollutants. Cauca River receives both biodegradable and non-biodegradable pollutants. Low DO events breaks the balance of the ecosystem and impedes the self-purification process. The monitoring activities over the last years have shown this condition of low levels of OD in the Cauca River and the impact on its self-purification capacity. The monitoring activities also have identified the type of discharge (point and diffuse) and received wastewater (domestic, industrial, urban runoff, runoff from agricultural areas, etc.). In this study DO has been used as the main water quality indicator. Because of this, it is important to differentiate between contaminants that are biodegradable and their impact on DO, and others that are more recalcitrant like pesticides, fertilizers, heavy metals, etc., which may have other (eco-toxicological) impacts. The latter category of contaminants do not disappear from aquatic environments, but may be accumulated in the river sediment and may cause long term eco-toxicological effects, including food chain accumulation. For this type of contamination the best management options are provided under Step 1 of the 3-SSA (minimisation and prevention).

The Salvajina dam began operation in 1985. One of its objectives was Cauca River water quality improvement. This was expected by increasing its pollution dilution capacity and thereby, improving of its self-purification capacity (Sandoval et al., 2007). However, this goal was not achieved, because: 1) the daily minimum flows are similar to the typical flow of dry season pre-Salvajina period (before 1985); 2) effective WWTP coverage of domestic and industrial wastewater is low; 3) diffuse pollution (urban and rural areas) has increased with the waterproofing of the cities, poor agricultural practices and the progressive deterioration of the watersheds. For the Cauca River, most of the self-purification capacity (Step 3 of Three Steps

Strategy Approach 3-SSA) was lost in the last 60 years. A wetland area of 300 km2 existing in the 50s was reduced in 1986 by 90% (Muñoz, 2012). This has significantly reduced the possible impact of interventions under Step 3 of the 3-SSA. Also for Step 1 and 2 of the 3-SSA there has not been significant progress. Minimization and prevention (Step 1) could be an effective strategy for pollution control in both, rural and urban environments. Also interventions under Step 3 could be further improved. This can be achieved by implementing sustainable urban drainage systems (SUDS) and modifying the operating criteria of the Salvajina reservoir to effectively contribute to the improvement of the Cauca river water quality. While these types of solutions are being developed, it is necessary to strengthen the early warning networks to manage the impact of disruptions in the drinking water service in the Cali city.

The issues associated with the 3-SSA are now mentioned in Colombian regulations and national policy documents (MAVDT, 2010). However, its implementation, as an integral strategy has been limited because the basin is not the 'unit of analysis', there is a limited institutional coordination, and there is weakness in the institutions responsible for improving the waters resource quality.

6.4 Conclusions

The Cauca River (La Balsa - Anacaro stretch) has a dynamic behaviour associated with the operation of the Salvajina reservoir, located 27.4 km upstream from the La Balsa Station and 139 km upstream from the Juanchito Station. This latter station was used as reference station because it is located a few meters from of the intake of Cali city's water supply system. In addition, the Cauca River receives pollutant discharges with dynamic behaviour, even during the dry season, including typical variations of discharges from WWTP of Cali city (WWTP-C), including pollutant peaks by out-of-operation periods of WWTP-C. However, one of the most critical dynamic situations is the pollutant flush happening during rainfall events (first flush effect), associated with diffuse and accumulated pollution in the urban and rural sectors of the South Drainage System of the Cali city.

The dynamic behaviour of water bodies and the pollution sources significantly affect the self-purification capacity of the water bodies (Step 3 of 3-SSA). Flow changes and pollution peaks generate variations in the dilution capacity and DO levels. In the Cauca River, when the pollution peaks coincide with periods of low flow, the minimum DO and self-purification capacity (Step 3 of 3-SSA) was reduced and anoxic water conditions reach the upstream point of water intake for Cali city.

The pulsed regime effect of the Salvajina reservoir on the hydraulic behaviour of the Cauca River and its impact on the water quality and self-purification capacity must be studied under dynamic conditions. For an average flow of 143 m^3/s at Juanchito Station and a steady-state (base line) condition, the Cauca River would remain under aerobic conditions, whereas in a dynamic flow condition (base line) a DO value close to zero is expected at Puerto Isaac Station.

The pollution associated to rainfall events (first flush effect) in the south drainage system of the city of Cali (South Channel), Scenarios S3 and S4, generated sharp reductions in DO concentrations at the water intake point. Values below 1.0 mg/L were found at this point, which results in frequent closures of the potable water purification plant. The pollution impact due to rainfall events is less critical when the peak of pollution was generated during daytime, when the Cauca river flows are higher. This condition also shows the effect in the water quality of the Cauca River as a source of water supply for the city of Cali due to the reservoir operation.

Step 3 of the 3-SSA can play an important role in cost-effective water quality management of rivers and associated water bodies. To study this, we must use dynamic modelling as a tool and consider the basin as a unit of analysis. Further dynamic conditions analysis for water quality studies in the Colombian rivers need to be implemented within the legislation and regulatory requirements, especially for those with a dynamic activity such as the Cauca River.

6.5 References

Abbott, J., Davies, P., Simkins, P., Morgan, C., Levin, D. and Robinson, P. (2013). Creating water sensitive places - scoping the potential for Water Sensitive Urban Design in the UK. CIRIA, London, UK.

CVC (2013). Cauca River. Daily report of levels and flows. Environmental Technical Direction. Environmental Information Systems. Corporación Autónoma Regional del Valle del Cauca CVC, Cali. (In Spanish)

EMCALI (2013). Flow and loads discharged to the Cauca river by the city of Cali. Empresas Municipales de Cali. (In Spanish)

EPSA (2014). Hourly data of turbine flow, power generation and water level of Salvajina reservoir, from October 2012 to December 2013. Empresa de Energía del Pacífico S.A. E.S.P. (In Spanish)

Galvis, A. (1988). Water quality simulation of the Cauca River. Calibration, verification and application. MSc Thesis in System and Industrial Engineering, Universidad del Valle, Cali, Colombia. (In Spanish)

Gijzen, H.J. (2006). The role of natural systems in urban water management in the City of the Future - a 3 Step Strategic Approach. *Ecohydrology and Hydrobiology*, 6(1-4), pp. 115-122.

González, S., López-Roldán, R., & Cortina, J.L. (2012). Presence and biological effects of emerging contaminants in Llobregat river basin: A review. *Environmental Pollution*, 161(2012), pp 83-92.

Hurtado, I.C. (2014) Efecto del embalse de Salvajina en la capacidad de autodepuración del río Cauca, MSc Thesis, Universidad del Valle, Cali, Colombia.

IDEAM (2011). Technical guidelines for national modelling of water resources. Water quality modelling in streams and reservoirs, Instituto de Hidrología, Meteorología y Estudios Ambientales, Ministerio de Ambiente, Vivienda y Desarrollo Territorial, República de Colombia. (In Spanish)

Ifabiyi, I.P. (2008). Self-purification of a freshwater stream in Ile-Ife: lessons for water management. *Human Ecology*, 24(2), pp 131-137.

Khan, S. and Singh, S.K. (2013). Assessment of the Impacts of Point Load on river Yamuna at Delhi Stretch, by DO-BOD Modelling of river, Using MATLAB Programming. *Engineering and Innovative Technology*, 2(10), pp 282-290.

MAVDT (Ministerio de Ambiente Vivienda y Desarrolllo Teritorial) (2010). National Policy for Integrated Water Resources Management. Deputy Minister of Environment. Ecosystem Management. Water Resources Group. Bogotá D.C., Colombia. (In Spanish)

Moreno G.E. (2014). Environmental analysis of the Upper Cauca river basin, its main actors and its impact on water supply of Cali city. What should we do?. MSc thesis in Industrial Engineering, Universidad ICESI, Cali, Colombia. (In Spanish).

Muñoz, E. (2012). Wetlands in the American continent. Concepts and classification in Peña, E.J., Cantera, J.R., Muñoz, E. (Eds.) Evaluation of pollution in aquatic ecosystems. Case study in the Sonso Lagoon, Upper Cauca river basin (pp. 23-49). Universidad Autonoma de Occidente and Universidad del Valle, Editorial Program, Cali, Colombia. (In Spanish)

Ostroumov, S.A. 2008 Basics of the molecular-ecological mechanism of water quality Formation and water self-Purification, *Contemporary Problems of Ecology*, 1(1), pp. 147-152.

Ramírez, C.A., Santacruz S., Bocanegra, R.A. and Sandoval, M.C. (2010). Salvajina reservoir Incidence on the flow regime of the Cauca river in its upper valley. *Ingeniería de los Recursos Naturales y del Ambiente*, 9 (2010), pp 89-99. (In Spanish)

Sandoval M.C., Ramirez, C.A. and Santacruz S. (2007). Monthly operating rule optimisation of Salvajina reservoir. *Ingeniería de los Recursos Naturales y del Ambiente*, 6(2007), pp 93-104. (In Spanish).

Streeter, H.W. and Phelps, E.B. (1925). A study of the pollution and natural purification of the Ohio river. Vol. III, Public Health Bulletin, N° 146, U.S. Public Health Service.

Universidad del Valle and Corporación Autonoma Regional del Valle del Cauca CVC (2007). Optimization of the water quality simulation model of the Cauca river. La Balsa - Anacaro strecht. Project report Volume XIII. Cauca river Modelling Project (PMC), Phase III. Cali, Colombia. (In Spanish)

Universidad del Valle and Corporación Autonoma Regional del Valle del Cauca CVC (2004). Sampling campaign for calibration purpose of Cauca River water quality model Project report Volume VI, Cali, Colombia. (In Spanish)

Universidad del Valle & Corporación Autonoma Regional del Valle del Cauca CVC (2009). Water quality modelling scenarios of the Cauca river. Project report, Cali, Colombia. (In Spanish).

Universidad del Valle and Empresas Publicas Municipales de Cali EMCALI (2006) Impact assessment of the proposed strategies by EMCALI for the management of wastewaters in the Cali city on the Cauca River water quality, Project Report, Cali, Colombia. (In Spanish).

Vagnetti R., Miana P., Fabris M. and Pavoni B. (2003). Self-purification ability of a resurgence stream, *Chemosphere*, 52(2003), pp 1781-1795.

Velez C.A., Alfonso L., Sanchez A., Galvis A. and Sepulveda G. (2014) Centinela: an early warning system for the water quality of the Cauca River. *Journal of Hydroinformatics*. 16(6), 1409-1424.

Von Sperling, M. (2005) Introduction to water quality and sewage treatment, 3th edition, Department of Sanitary and Environmental Engineering – Federal University of Mina Gerais, Belo Horizonte, Brazil. (In Portuguese).

Chapter 7
Validation of the Three-Step Strategic Approach for improving urban water management and water resource quality improvement

Source: CVC photo file

This chapter was submitted is based on:
Galvis. A., Van der Steen, N.P., and Gijzen, H.J. (2018) Validation of the Three-Step Strategic Approach for improving urban water management and water resource quality. *Water, MDPI Journal*, 10, 188; doi:10.3390/w10020188

Abstract

The impact on water resources caused by municipal wastewater discharges has become a critical and ever-growing environmental and public health concern. So far, interventions have been positioned largely 'at the end of the pipe', via the introduction of high-tech and innovative wastewater treatment technologies. This approach is incomplete, inefficient and expensive, and will not be able to address the rapidly growing global wastewater challenge. In order to be able to efficiently address this problem, it is important to adopt an integrated approach like the 3-Step Strategic Approach (3-SSA) consisting of: 1) minimization and prevention, 2) treatment for reuse and 3) stimulated natural self-purification. In this study, the 3-SSA was validated by applying it to the Upper Cauca river basin, in Colombia and comparing it to a Conventional Strategy. The pollutant load removed was 64,805 kg/d Biochemical Oxygen Demand BOD_5 (46%) for the conventional strategy and 69,402 kg/d BOD_5 (50%) for the unconventional strategy. Cost benefit analysis results clearly favoured the 3-SSA (unconventional strategy): NPV for the conventional strategy = $-276,318 \times 10^3$ Euros, and NPV for the unconventional strategy (3-SSA) = $+338,266 \times 10^3$ Euros. The application of the 3-SSA resulted in avoided costs for initial investments and operation and maintenance O&M, especially for groundwater wells and associated pumps for sugar cane irrigation. Furthermore, costs were avoided by optimization of wastewater treatment plants WWTPs, tariffs and by replacement of fertilizers.

7.1 Introduction

In order to be able to efficiently address problems caused by municipal wastewater discharge, it is important to adopt an integrated approach that includes control of contamination at source, followed by treatment and reuse, or responsible discharge of the final effluent. These 'cleaner production' principles have been successfully applied in the industrial sector and now these concepts are being applied to integrated water resources management. In this context, the conceptual model of the Three-Step Strategic Approach (3-SSA) was developed, consisting of: 1) minimization and prevention, 2) treatment for reuse and 3) stimulated natural self-purification (Gijzen, 2006; Galvis *et al.*, 2014a).

The minimization and prevention concept refer to the reduction of residues, emissions and discharges of any production process through measures that make it possible to decrease, to economically and technically feasible levels, the amount of contaminants generated which require treatment or final disposal (Cardona, 2007). However, the approach should go beyond only reducing emissions, by also looking at ways to reduce the use of raw materials (e.g. drinking water in this case) (Gijzen, 2001). Since the amount of personal human waste (urine and faecal matter) will remain the same, by using less water more concentrated wastewater is produced, which lends itself better for treatment in the direction of reuse. The minimization proposals can be classified in three main actions (Nhapi and Gijzen, 2005; Cardona, 2007): a) reduction at source, which includes a change in consumption habits and application of low water consumption devices; b) in situ recycling techniques, and c) rainwater harvesting. The first action proposes a shift to low consumption devices, such as water-saving toilets, showers and aired faucets that generate a decrease in the consumption of water, allowing for the possibility of supplying more users, without the need for new water sources and treatment capacity. The second and third actions, in situ recycling techniques, recognize new alternative water sources, such as rainwater harvesting and grey water. Lastly, the use of treated grey water is feasible for toilet flushing, plant watering, and the washing of floors and outdoor areas (Mejia *et al.*, 2004; Gijzen, 2006; Liu *et al.*, 2010), as well as golf courses, agriculture and groundwater recharge (Ottoson and Stenström, 2003).

Water reuse refers to utilization of water previously used one or more times in some activities to satisfy the needs of other uses, including the original. Reuse requires the processing of municipal wastewater to achieve specific quality criteria suitable for subsequent (re-)use (Asano *et al.*, 2007; U.S. EPA, 2012). Treated wastewater may be used beneficially in activities such as crop irrigation, industrial processes, cleaning or washing activities, protection of water resources, prevention of pollution, recovery of water and nutrients for agriculture, savings in freshwater use and wastewater treatment costs, etc. (Capra and Scicolone, 2007). Besides, wastewater reuse as an additional source of water represents dual environmental benefits due to the decrease in the amount of water used for sensitive ecosystem, recreational activities and a decrease in wastewater discharges, leading to a reduction/prevention of water resource contamination (U.S. EPA, 1998). To meet current and future reclamation requirements and regulations, the selection of technologies for water reuse will involve careful consideration and evaluation of numerous factors. On selecting technologies for water reuse, consideration has to be given as to whether existing facilities are to be modified or upgraded, or an entirely new facility is to be constructed. In general, both physical and operational factors will have to be considered (Asano *et al.*, 2007). The process can start with a pre-selection where technologies considered should ensure the production of an effluent that meets: 1) the quality requirements for the type of reuse considered, or 2) local discharge criteria. Based on this, it will be necessary to choose the most appropriate wastewater treatment alternative, considering the technical, social, environmental and economic issues.

Discharges that are not avoided via prevention/minimization (Step 1) and reuse of treated effluents (Step 2) will be discharged to water bodies. At this stage, the 3-SSA proposes to consider interventions that maximize the self-purification (natural or stimulated) capacity of receiving water bodies (Step 3). When a river is polluted, the water quality deteriorates, limiting water use and ecosystem functions (González *et al.*, 2012). However, the self-purification capacity of a river allows it to restore (partially or fully) its quality through re-aeration and natural processes of biodegradation (Von Sperling, 2005). The mechanisms of self-purification can be in the form of dilution of polluted water with an influx of surface or groundwater or through certain complex hydrological, micro-biological and chemical processes (Ifabiyi, 2008; Ostroumov, 2008). Under Step 3 measures can be introduced that stimulate the 'self-purification' capacity of a water resource, for instance by introducing ecohydrology interventions. Since anoxic water bodies generally have lower self-purification capacity, DO concentration is a primary measure of a stream's health; it responds to the biochemical oxygen demand (BOD) load (Khan and Singh, 2013). This is why oxygen demand (DO) has been traditionally used to assess the pollution degree and self-purification capacity of water bodies. DO can be easily measured; however, the complex mechanisms involved in DO must be studied by mathematical modelling (Von Sperling, 2005). Streeter and Phelps developed the first models in 1925. They developed a balance between the dissolved oxygen supply rate from re-aeration and the dissolved oxygen consumption rate from stabilization of an organic waste in which the biochemical oxygen demand (BOD) de-oxygenation rate was expressed as an empirical first order reaction, producing the classic dissolved oxygen (DO) sag model. This model is usually studied through mathematical modelling, either for steady state or for dynamic conditions. The selection of the model will depend on the objectives of the study, the specific characteristics of the study site and the availability of information (Galvis *et al.*, 2014b).

In this study the 3-SSA (non-conventional strategy) was validated by applying it to the Upper Cauca river basin in Colombia and comparing it to a Conventional Strategy, which considers a 'business as usual scenario' of high water use, end-of-pipe wastewater treatment and conventional water supply providing drinking water quality for all uses. The Cauca River is the second most important fluvial artery of Colombia and the main hydric source of the Colombian southwest. Although actions aimed at pollution control in the Upper Cauca river basin date back over 40 years, the river water quality in the study area continues to deteriorate. This situation persists despite the fact that 19 of the 41 municipalities have installed WWTPs. In this research the Unconventional Strategy includes reduction in water consumption and reuse of treated wastewater in households and for sugarcane crop irrigation. It also considers prioritization of investments to maximize impact in improving the water quality of the Cauca River in the study area, targeting interventions in watersheds and municipalities with the highest pollutant load and located upstream of the river segments with the lowest DO. This study defines a Baseline (2013, dry season condition) and scenarios for Conventional and Unconventional strategies towards 2033. The MIKE 11 model was used to study BOD_5 and DO behaviour in the Cauca River for each strategy. Additionally, the strategies were compared using cost benefit analysis (CBA) (Brent, 2006). This study uses the incremental cost-benefit analysis and it does not consider the common costs and benefits to compare the strategies (Bos *et al.*, 2004).

7.2 Methods

7.2.1 Study area

The study area is the Upper Cauca river basin (Figure 7.1), in particular the stretch from La Balsa km 27.4 (980.52 meters above sea level m.a.s.l) to Anacaro km 416.1 (895.56 m.a.s.l). The Cauca River is the main water resource of the Colombian southwest. It has a total longitude of 1,204 km with a tributary area of 59,074 km^2. The La Balsa -Anacaro stretch has an average width of 105 m and the depth varies between 3.5 and 8.0 m. The longitudinal profile of the Cauca river shows a concave shape with a hydraulic slope, which oscillates between 1.5×10^{-4} m/m and 7×10^{-4} m/m (Ramirez *et al.*, 2010). The average annual rainfall varies between 938 mm (the central sector) and 1,882 mm (southern sector). There are two dry season periods: December - February and June-September. Rainy days per year vary between 100 days (central sector) and 133 days (northern sector) (Sandoval and Ramirez, 2007). The sugar cane crops and the Colombian sugar industry are located in the flat area along the Upper Cauca river basin. In the mountain area, there are coffee crops and associated industry. The Cauca River is used for fishing, recreation, power generation, riverbed matter extraction, irrigation, industry, and as a main source for drinking water supply. The Salvajina reservoir began operations in 1985 and is part of the regulation project of the Cauca River, implemented for flood control, improving water quality and power generation. The reservoir operates with a minimum flow discharge of 60 m^3/s and average daily flow rate of 143 m^3/s in the Juanchito station (Sandoval *et al.*, 2007). The Cauca River is also used as a receiving source for solid waste and dumping of industrial and domestic wastewater, which is contributing to the decline in water quality.

7.2.2 Baseline conditions-2013

The baseline conditions correspond to the dry season of 2013. In that year the study area had 3.8 million inhabitants. For these conditions, the Cauca River received approximately 140 T/d of BOD_5 in the La Balsa - Anacaro Stretch. Municipality of Cali (rivers and urban area) and four other prioritized sub-catchments), located upstream of the minimum DO station, represent

70.3% of the total pollutant load (BOD₅) discharged throughout the study stretch from pollution point sources (see Figure 7.1 and Table 7.1). The main characteristics of the baseline condition are described below.

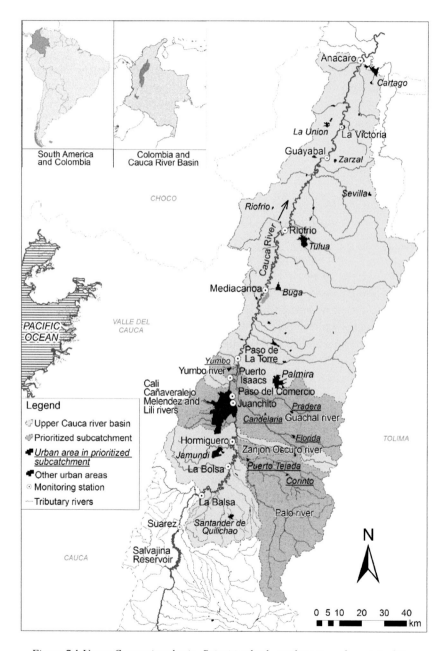

Figure 7.1 Upper Cauca river basin. Prioritized sub-catchments and municipalities

Table 7.1. BOD$_5$ discharged to the Cauca River in the La Balsa-Anacaro Stretch. Baseline. Conditions 2013 for the dry seasons and mean flow of 143 m^3/s, at Juanchito Station

Prioritized	BOD$_5$	Percentage %	
sub-catchment	(T/d)	Partial (1)	Accumulated (2)
1. Cali, Cañaveralejo, Melendez and Lili rivers + urban area of Cali city	72.2	51.6	51.6
2. Palo River	7.6	5.4	57.0
3. Zanjon Oscuro River	7.1	5.1	62.1
4. Guachal River	7.0	5.0	67.1
5. Yumbo River	4.4	3.2	70.3
Other discharges	41.5	29.7	100
Total	139.8	100	

Most of the wastewater discharges of Cali city originate in the urban sub-catchments of Cali, the Lili, Melendez and Cañaveralejo rivers. These three rivers flow into the Cali sewerage system via the South Channel (982 L/s, BOD$_5$: 2.4 T/d), while the effluent of the WWTP of Cali city (6,720 L/s, BOD$_5$: 61.4 T/d) discharges to the Cauca River. Another part of the wastewater of the urban area of Cali is discharged directly to the Cauca River via two pumping stations: Floralia (212 L/s; BOD$_5$: 3.5 T/d) and Puerto Mallarino (842 L/s; BOD$_5$: 4.9 T/d).

In the 31 sub-catchments of the study area there are 38 municipalities. For the Baseline Condition 19 municipalities had WWTP, 4 of which were out of operation (municipalities: Villa Rica, Pradera, Yumbo and Cerrito). The WWTP technologies for Baseline Conditions were: 1) preliminary treatment + Up-flow Anaerobic Sludge Blanket (UASB) + trickling filter+ secondary settler (two municipalities, flow: 30 - 300 L/s); 2) preliminary treatment + anaerobic pond+ facultative pond (six municipalities, flow: 30 - 80 L/s); 3) preliminary treatment+ high rate anaerobic pond + facultative pond (municipality of Cerrito: 90 L/s); 4) preliminary treatment + high rate trickling filter + secondary clarifier (municipality of Tulua: 330 L/s); 5) septic tank + upflow anaerobic filter (six municipalities, flow: 2 - 50 L/s); 6) preliminary treatment + anaerobic pond + aerobic filter (two municipalities, flow: 15 - 25 L/s); 7) preliminary treatment + Dissolved Air Flotation unit (DAF) (municipality of Yumbo: 60 L/s). Sludge drying beds are used in most cases for sludge handling.

The infrastructure corresponding to Baseline (2013) of the WWTP of Yumbo was completely disregarded, because the system was not in operation. On the other hand, the Villa Rica WWTP needs to be optimized and 4 WWTPs (municipalities: Guachené, Miranda, El Cerrito and La Union) need to be expanded to ensure the required treatment level. In the Upper Cauca river basin, there are three municipalities, each with discharges to two different sub-catchments: Puerto Tejada (Palo and Zanjon Oscuro rivers); El Cerrito (Cerrito and Zabaletas rivers) and Ginebra (Zabaletas and Guabas rivers). For each of these municipalities two WWTPs were considered.

In larger municipalities, especially in Cali city, industries with direct discharges to the municipal sewer system were included. This load was estimated at 6.7 T/d BOD_5. However, in the study area (Upper Cauca river basin) there were, for Baseline Conditions, over 100 industries, most of them with treatment plants whose effluent was discharged into the Cauca River directly or through its tributaries. These discharges accounted for approximately 25 T/d BOD_5. 80% of this load corresponded to only 12 industries, which had relatively high BOD_5 discharges despite the fact that these industries had wastewater treatment plants.

7.2.3 Formulating Strategies: Conventional ('business as usual scenario') and Unconventional (3-SSA)

Two types of strategies projected to 2033 were defined in the context of the Upper Cauca river basin, La Balsa-Anacaro stretch: 1) Conventional Strategy, which considers a 'business as usual scenario' of high water use, end-of-pipe wastewater treatment and conventional water supply providing drinking water quality for all uses; 2) Unconventional Strategy, applying the 3-SSA. For the two strategies, industrial discharges to the Cauca River (directly or via tributaries) remain constant over the horizon of the analysis. The projection of pollutant loads was performed to the projection horizon (2033). The construction of the infrastructure was completed in 2016. For the Baseline Conditions (2013) the consolidated area of Cali city had 1.85 million inhabitants, distributed over 74% single family housing units and 26% multifamily housing blocks (Alcaldía de Santiago de Cali, 2013). In this research the existing urban area for the Baseline Conditions is considered to be a 'consolidated area'. It is assumed that the population of the 'consolidated area' remains constant until 2033. The future population growth will be accommodated in the expansion area of Cali (607,696 inhabitants in 2033) with a distribution of 85% single family housing and 15% in multifamily housing. For the other municipalities only single family homes will be scheduled.

Available information in the Public Services Unified Information System of the Republic of Colombia (SUI) and the National Administrative Department of Statistics of Colombia Republic (DANE) was used for the construction of the Baseline Conditions (2013). Information provided by the environmental authorities in the region (CVC and CRC), Research Centre of Sugarcane Cenicaña (Cruz, 2015), municipalities, consultant companies and service providers were used to characterize the hydrology, the hydraulics and the water quality for the Cauca River (main channel), tributaries and sub-catchments. Some of this information has been systematized and analysed in previous reports (Universidad del Valle and EMCALI, 2006; Universidad del Valle and CVC, 2007 and 2009). The projection of population growth was made based on (DANE, 2005). For the Cali city case the average flow of wastewater was estimated as 80% of the water supply and the maximum hourly wastewater flow factor (FM) was obtained from the expression $FM=2.3(Q_m)^{-0.062}$, where Q_m is average flow (EMCALI, 1999). For both the Conventional and Unconventional strategies, it was assumed that the flow rates and BOD_5 loads from the industrial sectors, located outside of the urban areas, remain constant and equal to the baseline values, for the entire projection horizon (2033). The same assumption was made for the wastewater produced by the scattered settlements.

Conventional Strategy
For the Conventional Strategy, in 2016 (i.e. 3 years after the baseline year) all municipalities were assumed to have a WWTP that ensures compliance with existing national regulations, reaching 80% removal of BOD_5 and TSS planning horizon. This involves optimizing/extension of existing WWTPs (Baseline Conditions) and building new WWTPs for all the municipalities in the study area that still had no WWTP. With regard to Cali it is assumed that the future population will settle in the expansion area and a second WWTP will be built there. For technology selection of these new plants, information of existing plants was compiled and cost models for major technological schemes were developed.

For the Conventional Strategy most of Cali's wastewater discharges reach the existing treatment plant (WWTP-C) and is treated at advanced primary level for Baseline Conditions. An activated sludge step-feed system was selected, according to Hazen and Sawyer's design for Cali city, with the following characteristics: flow 7,396 L/s, BOD_5 influent: 110.8 T/d; BOD_5 effluent: 22.2 T/d. Cali city will have an additional treatment plant, which will receive wastewater from the expansion area and South Channel illegal connections. The selected technological scheme includes: primary treatment + UASB + maturation pond, flow: 1,075 L/s; BOD_5 influent: 27.6 T/d and BOD_5 effluent: 5.5 T/d. In addition to these two new plants for Cali, 24 WWTPs for other municipalities are required. These plants were distributed as follows: 19 in municipalities that do not have WWTPs and according to their topographic features require only one treatment plant; two municipalities (Cerrito and Ginebra) requiring each an additional treatment plant to the one already existing, and the municipality of Puerto Tejada which requires two treatment plants. Additionally, there is the WWTP of Yumbo whose existing infrastructure was completely disregarded for this analysis. The technologies for these 24 WWTPs for the Conventional Strategy were:

Scheme 1: Advanced primary (existing) + activated sludge step feed (Cali, WWTP-C)
Scheme 2: Preliminary + UASB + maturation pond (Cali, expansion area)
Scheme 3: Preliminary + UASB + trickling filter + secondary settler (municipalities: Santander de Quilichao, Sevilla, Zarzal)
Scheme 4: Preliminary + anaerobic pond + facultative pond (municipalities: Puerto Tejada, discharge to Zanjon Oscuro River, Candelaria, Yotoco, El Cerrito, San Pedro, Andalucía, Vijes, Bugalagrande, Ansermanuevo, La Victoria, Obando)
Scheme 5: Preliminary + high rate trickling filter + secondary clarifier (municipalities: Jamundí, Yumbo, Palmira)
Scheme 6: Septic tank + anaerobic upflow filter (municipalities: Puerto Tejada, discharge to Palo River, Ginebra, Trujillo)
Scheme 7: Preliminary + UASB + facultative pond (Buga)
Scheme 8: Preliminary + anaerobic pond + aerobic filter (Bolivar)

Scheme 9: Preliminary + high rate anaerobic pond + facultative pond (Florida)

With regard to sludge handling, the following technologies were used: for Scheme 1, thermal and for Scheme 5, sludge thickener + primary sludge digester + secondary sludge digester. For other schemes drying beds were selected.

It is important to take account that in this research, conventional or non-conventional does not refer to the technology of WWTPs, but the strategy. The technology indicated here was based on existing WWTPs and technologies that have been considered in preliminary studies and designs by consultants for the municipalities. The Table 7.2 shows the BOD_5 discharged to the Cauca River for the Conventional Strategy, projected to 2033.

Unconventional Strategy
For the Unconventional Strategy, the 3-SSA was applied in prioritized sub-catchments and municipalities from 2016 onwards prioritised sub-catchments, assuming required infrastructure and operational measures were fully in place by 2016. However, in Step 1 (prevention and minimization), low consumption devices, rainwater harvesting and grey water reuse were applied, along the project horizon, in major urban centres, with different criteria for existing households and new households. Step 2 (treatment for reuse) includes the reuse of WWTP effluent for agricultural irrigation. Step 3 (stimulate natural self-purification) identifies the sub-catchments with the highest contribution of pollutant load (BOD_5) and prioritizing interventions of steps 1 and 2 upstream of the Paso de La Torre Station (Figure 7.1), where the minimum DO occur (Baseline Conditions).

Step 1: Prevention and minimization
This includes reduction in water consumption, by changing habits, use of low consumption devices, grey water reuse and rainwater harvesting (Galvis *et al.,* 2014a). With the implementation of Step 1, the average consumption was reduced with different criteria for multifamily households and single-family households. For new multifamily dwellings (in Cali city only) a small reduction of BOD_5 and TSS loads via the grey water reuse was assumed. The unit consumption for the Baseline Conditions (2013) were: Cali consolidated area, including drainage area of WWTP-C: 250 L/p/d and expansion area: 150 L/p/d (proposed for Conventional Strategy), while consumption in the prioritized municipalities varied between 114 and 184 L/s. As a result of the strategies for minimizing these consumptions, they were reduced to 95.3 L/s for Cali consolidate area, 69.3 L/s for the expansion area of Cali and 93 L/s for the other prioritized municipalities. The greatest reduction in consumption for the expansion area of Cali was because in new multifamily households, besides the change of habits and implementing low consumption devices, reuse of treated greywater and rainwater harvesting were included. In this case the pollutant load reduction was small (1.1 T/d BOD_5), because the prevention measures implemented in the multifamily housing in the expansion area of Cali were marginal, so only 5% of BOD_5 prevention was achieved. However, the sewage was more concentrated due to lower water consumption. Among the largest cities in Colombia, Cali is the

one with the highest water consumption. The defined value here is 11.3 m³/household/month for single households and 11.8 m³/household/month for multifamily housing. However, for the expansion area of Cali, the change of habits, low consumption devices, combined with grey water reuse and rainwater harvesting reduces consumption to 7.9 m³/household/month in multifamily households. With these approaches applied to all the prioritized municipalities, a total reduction in consumption of 5,098 L/s is achieved, which also leads to reduction in wastewater flows. This represents benefits by the following avoided costs: water and sewer tariffs, tax for water use in the water supply system, tax for wastewater discharges directly to water bodies and smaller infrastructure of water supply systems (water supply network and drinking water treatment plant DWTP) and WWTP.

Step 2: Treatment for reuse
This step includes the reuse of treated wastewater in the irrigation of sugarcane crops and prioritization of investments to maximize impact in improving the water quality of the Cauca River in the study area, targeting interventions in municipalities and sub-catchments with the highest pollutant load. In 2016, to ensure compliance with national regulations (removal of BOD and TSS), the prioritized municipalities had to guarantee the quality of WWTP effluent for irrigation of existing sugar cane crops. Technology was selected involving public health criteria (WHO, 2006) and agrological quality for agriculture irrigation (Ayers and Wescot, 1987). This involved the optimization of existing plants and building new WWTPs in prioritized municipalities. To analyse the reuse feasibility, it was necessary to study the aquifer vulnerability (Foster and Skinner, 1995) and to calculate the required irrigation area via cartographic analysis using ArcGIS 9.3. To complete this analysis, it was also necessary to develop the agricultural plan to verify the projected sugar cane crops water demand, developing a simplified water balance (Sokolov and Chapman, 1981), including the calculation of: effective precipitation (Doorenbos and Kassam, 1979), current evapotranspiration using the Food Agriculture Organization (FAO) methodology combined (Allen *et al.*, 2006) and Cenicaña (Torres and Carbonel, 1996). Irrigation is by furrows with efficiency of about 40% (Diaz, 2006). This means that of every 100 L that are used in the irrigation of crops, only 40 L are actually used by the crop.

With the minimizing of consumption, influent flow to WWTP-C is reduced from 7,396 to 4,167 L/s, while the BOD_5 load (T/d) remains the same as for the Conventional Strategy. Approximately 80% (3,326 L/s) of the total flow of WWTP effluent was used to irrigate sugar cane crops located on the right bank of the Cauca River. The remaining flow (841 L/s) was discharged directly to the Cauca River, considering that the removal of 80% in BOD_5 (T/d) and TSS (T/d) was achieved (Colombian regulations in 2013), without the need to build another treatment plant for this flow. The technology used to guarantee water quality for irrigation consisted of the following processes: advanced primary treatment (Baseline Conditions) + UASB + maturation pond + maturation pond. For the expansion area of Cali, prevention and minimization strategies reduced the influent flow to the WWTP to 576 L/s and to ensure the quality of the effluent for agricultural reuse, a maturation pond was added. According to the irrigation area characteristics and the agricultural plan, it was possible to irrigate 3,080 ha during 334 days per year with the effluent of WWTP-C and to irrigate 2,276 ha of sugar cane crops during 62 days per year with the effluent of the WWTP of the expansion area. During agricultural irrigation days with treated wastewater, two direct wastewater discharges into the Cauca River were avoided: 10.9 T/d from WWTP-C and 0.6 T/d BOD_5 from the WWTP of the expansion area. In the cases of Puerto Tejada WWTP (discharging effluent into the Zanjón Oscuro River) and Candelaria WWTP, selected technology in the Conventional Strategy guaranteed the water quality of effluent for reuse, so for this case the implementation of any additional process was not required. For the municipality of Florida, reuse of WWTP effluent was not feasible due to the vulnerability of the aquifers. The local environmental authority, based on (Foster and Skinner, 1995), has defined this vulnerability. It is a function of depth water table, net recharge, aquifer media, media soil, topography, impact of vadose zone, hydraulic conductivity, ground water occurrence and fertilization with nitrogen. The other prioritized municipalities (Corinto, Puerto Tejada, Yumbo, Candelaria, Pradera and Palmira) corresponded to 705 L/s of wastewater for reuse in sugar cane crops, in 2033. To ensure water quality for reuse, it was necessary, in each municipality, to add a maturation pond to the selected technological scheme of the Conventional Strategy, to meet the standards of pathogen removal, where helminth eggs is a critical parameter. According to the irrigated area characteristics and results of the agricultural plan, it was possible to irrigate 937 ha during 304 days in the municipality of Yumbo. For the remaining municipalities, considered together, it was possible to irrigate 3,332 ha during 62 days per year.

In summary, for the Unconventional Strategy, steps 1 and 2 were implemented only for Cali and municipalities of greater contribution of pollutant load in the prioritized sub-catchments: the Palo River (municipalities: Corinto and Puerto Tejada); the Zanjon Oscuro River (municipality: Puerto Tejada); the Yumbo River (municipality: Yumbo); the Guachal River (municipalities: Candelaria, Palmira and Pradera). The Table 7.2 shows the BOD_5 discharged to the Cauca River for the Unconventional Strategy (3-SSA), projected to 2033.

Step 3: Self-purification capacity. In this research, the stimulation of the self-purification capacity of the waterbody was associated with the prioritization of the investments in steps 1 and 2, upstream of the station with the minimum DO. This increases this minimum value and avoids the Cauca River to reach anaerobic conditions, which would limit the natural self-purification process. Table 7.2 shows the BOD_5 discharges for each strategy. For the Conventional Strategy in 2033 the total load discharged to the Cauca River was 75 T/d BOD_5, which is a reduction of 46.4% compared to the Baseline Conditions (2013) total discharge, while for the Unconventional Strategy the discharge was 70.5 T/d BOD_5 which means a reduction of 50%.

Table 7.2 BOD₅ discharges to the Cauca river in the La Balsa-Anacaro Stretch. Baseline 2013 for the dry season and 2033 projections for conventional and unconventional (3-SSA) strategies

Monitoring station	Abscissa (km)	Tributaries and discharges	BOD$_5$ discharged (kg/d)		
			Baseline 2013	Conventional Strategy 2033	Unconventional Strat. 2033 [1]
La Balsa	27.38	La Teta River	366	466	466
		Quinamayó River	643	261	1,024
		La Quebrada River	209	83	260
		Claro River	734	1,088	1,088
La Bolsa	78.86				
		Palo River	7,543	7,047	6,982
		Jamundí River	1,199	538	1,817
Hormiguero	112.82				
		Zanjón Oscuro River	7,122	4,246	3,471
		Desbaratado River	96	69	69
		WWTP- Expansion area of Cali	0	5,513	0
Antes de Navarro	127.00				
		South Channel	2,391	189	189
		P. Mallarino pumping station	4,887	0	0
Juanchito	139.02				
		WWTP -Cañaveralejo	61,420	22,156	13,676
		Cartones del Valle (industry)	0	0	0
Paso del Comercio	144.56				
		Empaques industriales (industry)	1,286	1,286	1,286
		Floralia pumping station	3,527	0	0
		Cali River	4,021	4,017	4,017
		Arroyohondo district	1,703	1,703	1,703
		Arroyohondo River	67	104	104
		Propal (industry)	267	267	267
		Puerto Isaacs (industrial district)	18	18	18
		Cencar (industry)	464	464	464
Puerto Isaacs	155.04				
		Acopi (industrial district)	1,104	1,104	1,104
		Cementos del Valle (industry)	11	11	11
		Smurfit-Cartón Colombia	3,206	3,206	3,206
		La Dolores (industrial district)	117	117	117
		Yumbo River	4,402	1,707	391
		Guachal River	7,037	4,027	3,127
Paso de la Torre	171.03				
		Amaime River	702	717	717
		Vijes River	10	10	10
		Municipality of Vijes	260	76	379
		Cerrito River	3,483	1,850	3,772
Vijes	186.54				
		Zabaletas River	495	265	422
		Guabas River	715	580	616
		Sonso River	384	390	390
		Yotoco River	94	24	113

Monitoring station	Abscissa (km)	Tributaries and discharges	BOD₅ discharged (kg/d)		
			Baseline 2013	Conventional Strategy 2033	Unconventi onal Strat. 2033 [1]
Yotoco	212.73				
		Mediacanoa River	30	30	30
Mediacanoa	220.92				
		Guadalajara River	191	170	170
		Piedras River	25	23	23
		Carmelita (sugar mill, industry)	147	147	147
		Burriga Channel	2,887	838	2,950
		Riofrio River	1,887	1,036	1,629
Ríofrio	284.77				
		Tuluá River	2,724	4,428	3,107
		Morales River	752	115	228
		Bugalagrande River	2,547	1,535	2,202
		Municipality of Bolívar	133	27	133
		La Paila River	5,190	0	0
		Municipality of Zarzal	1,189	1,988	5,050
		Municipality of Roldanillo	286	293	1,463
La Victoria	369.87				
		Municipality of Ansermanuevo	512	130	590
		Municipality of La Unión	511	428	491
		Municipality of La Victoria	387	99	387
		Municipality of Toro	48	48	48
Anacaro	416.06	Municipality of Obando	430	120	533
Total load (kg/d of BOD₅)			139,859	75,054	70,457

(1) In the table, some values of BOD₅ discharged (kg/d) to Cauca River for the Unconventional Strategy (3-SSA) 2033, are higher than corresponding to Conventional Strategy 2033, because the Unconventional Strategy was only applied in the prioritized sub-catchments.

7.2.4 Mathematical modelling to assess the impact of strategies on water quality of the Cauca River

The hydrodynamic and water quality model of the Cauca River was implemented in the MIKE 11 model. The Cauca River has 15 monitoring stations in the La Balsa–Anacaro stretch (Figure 1.1). The calibration and validation of the quantity (roughness) and quality (BOD, DO) components to apply the MIKE 11 model were based on (Universidad del Valle and CVC, 2007). The model consists of 387 cross sections, 2 external boundaries: La Balsa (km 27.4) and Anacaro (km 416.1), 95 internal boundaries which include 34 rivers and streams, municipal wastewater discharges, 24 industrial wastewater discharges and 36 water extraction sites (Galvis *et al.*, 2014b). Two monitoring campaigns were used: calibration (2005) and validation (2003).

The quality component of the MIKE 11 model at Level 1 and the Churchill equation for the re-aeration calculation were selected. Then the values resulting from the calibration-validation process are presented: Strickler roughness ($m^{1/3}$ s^{-1}); BOD_5 degradation constant (d^{-1}) and Benthic Oxygen Demand (g $O_2/m^2/d$). The values are presented in this order for each monitoring station on the Cauca River, the La Balsa–Anacaro stretch: La Balsa (40; 0.15; 1.5); La Bolsa (20; 0.15; 2) Hormiguero (40; 0.3; 3); Juanchito (33; 0.4; 5); Puerto Isaacs (60; 0.35; 5); Paso de la Torre (60; 0.33; 3); Mediacanoa (34; 0.2; 2); Guayabal (30; 0.17; 1); La Victoria (33; 0.17; 1) and Anacaro (32; 0.17; 1).

7.2.5 Cost Benefit Analysis (CBA)

Environmental and economic benefits were calculated. Common benefits, like health benefits, were not included, and only the incremental costs and incremental benefits were considered. 'Incremental' means that common benefits and common costs were not considered. Additionally, it means that there are differentiated costs and benefits only where the relative values between the two strategies were considered. For example, for prioritized municipalities of the Unconventional Strategy, additional costs were included for additional treatment processes to ensure the wastewater quality of the effluent of WWTP to irrigate sugar cane crops. For costing, constant prices were used without inflation (Boardman *et al.*, 2001). Infrastructure investments were projected to 20 years and a project horizon for the cost-benefit evaluation of 20 years was adopted. A social discount rate of 11.75% was applied (Comisión de Regulación de Agua Potable y Saneamiento Básico, 2013). For the Conventional Strategy the following was calculated: initial investment cost of the new WWTPs and optimization cost of existing WWTPs and operation and maintenance (O&M) of new and existing WWTPs. On the benefits side, reduction in tax for wastewater discharged to water bodies was calculated.

Initial investment and O&M cost associated with the Unconventional Strategy included: use of low consumption devices, rainwater harvesting, grey water reuse, optimization of WWTPs for reuse of the effluent, agricultural irrigation network and the pumping of the effluent of the Cali wastewater treatment (WWTP-C), to bring treated wastewater from the left bank to the right bank of the Cauca River, to reach sugarcane farms. The incremental benefits were corresponding to avoided cost due to implementation of 3-SSA. These benefits (avoided costs) have been classified into four groups: 1) initial investment and O&M of the drinking water distribution network, the WWTP and infrastructure (wells and pumping stations) for irrigation of sugar cane crops using groundwater; 2) reduction in water supply tariff and sewer tariff; 3) saving from reduced use of fertilizers and reduction in payment of fee for water use; 4) reduction in tax for wastewater discharge directly to water bodies. Information from local institutions and cost models obtained with information about the region (Sanchez, 2013) were used to obtain the initial investment and O&M costs of the WWTP. This same method was used to estimate the costs associated with the water supply infrastructure and wells and pumping stations for irrigation of sugarcane crops (Colpozos, 2010). The cost of power consumption was estimated as 0.13 €/kW-h.

In the CBA, Year 1 corresponds to Baseline Conditions (2013) and major infrastructure investments was proposed to be made in Year 3 (2016). Investments in grey water reuse and rainwater harvesting are done gradually between 2016 and 2033. The costs and benefits associated with O&M, taxes and fees were considered each year from Year 4 (2017) until Year 20 (2033). Costs were obtained in Colombian pesos and a conversion rate of 1 Euro (€) = 2,500 Colombian pesos was used. Based on information specific to sugar cane crops in the Valle del Cauca (Ministerio de Agricultura y Desarrollo Rural, 2010) the following prices for fertilizers were used: NPK = 0.53 €/kg and urea = 0.58 €/kg. For taxes, fees and tariffs specific values were applied to each prioritized municipality. The information was obtained from the local and regional environmental authorities (CVC, 2010 and 2012). The ranges corresponding to the Baseline Conditions were: water supply tariffs: 0.27 to 0.42 €/m³/month; sewerage tariffs: 0.16 to 0.49 €/m³/month; tax for surface water for domestic use: from 0.0003 to 0.0009 €/m³; tax for groundwater for agricultural irrigation use: 0.0003 to 0.001 €/m³; tax for wastewater discharges directly to water bodies: 0.020 €/kg SST and 0.047 €/kg BOD₅.

7.3 Results

7.3.1 Impact of strategies on water quality of the Cauca River

The MIKE 11 model showed that the minimum DO for the baseline was 0.6 mg/L (Puerto Isaacs Station, km 155) (Table 7.2), while the implementation of the Conventional and Unconventional strategies caused this value to increase to 1.6 mg/L and 2.0 mg/L (Paso de La Torre Station, km 171), respectively.

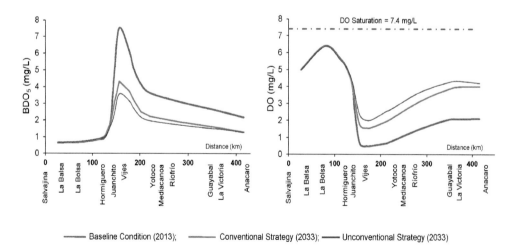

Figure 7.2 BOD₅ and DO profiles for the baseline (2013), Conventional and Unconventional strategies (2033). Data of dry season with average flow at Juanchito Station of 143 m³/s

7.3.2 Cost Benefit Analysis (CBA)

A CBA was performed based on incremental costs and benefits for the Conventional (Table 7.3 and Table 7.4) and the Unconventional Strategy, applying 3-SSA (Table 7.5 y Table 7.6). In all cases, the new treatment plants in the Unconventional Strategy had a lower net cost. The results show that the NPV (-276,318x10^3 €) is unfavourable for the Conventional Strategy. In contrast the NPV (+338,266x10^3 €) for the Unconventional Strategy (3-SSA) shows its advantage over the Conventional Strategy (Figure 7.3).

Table 7.3. Incremental cost of implementing the Conventional Strategy (thousands of €)

Item	NPV	Year 3	5	8	10	20
Initial investment						
- Secondary treatment WWTP-C and WWTP expansion area (Cali)	172,774	241,113				
- WWTP for other municipalities	21,874	30,526				
- Optimization of WWTP (municipalities)	3,134	4,374				
Operation and maintenance O&M						
- Secondary treatment WWTP-C and WWTP expansion area (Cali)	69,414		13,411	13,411	13,411	13,411
- WWTP other (municipalities)	10,030		1,956	1,956	1,956	1,956
- Optimization of WWTP (municipalities)	1,518		293	293	293	293
Total incremental cost	**278,744**	**27,6013**	**15,660**	**15,660**	**15,660**	**15,660**

Table 7.4. Incremental benefits of implementing the Conventional Strategy (thousands of €)

Item	NPV	Year 3	5	8	10	20
Reduction in tax for wastewater discharged directly to water bodies	2,426		457	471	482	539
Total incremental benefits	**2,426**		**457**	**471**	**482**	**539**

Table 7.5. Incremental cost of implementing the Unconventional Strategic 3-SSA (thousands of €)

Item	NPV	Year 3	5	8	10	20
Initial investment						
Low consumption devices+ rainwater harvesting + grey water reuse	1,171	185	190	197	201	227
Pumping station of treated WW (Cali) viaduct for transfer wastewater to the right side of river	1,806	2,521				
Water irrigation network for reuse (Cali, expansion area of Cali and other municipalities)	1,881	2,625				
Operation and maintenance O&M						
Rainwater harvesting + grey water reuse (Cali expansion area, multifamily households)	726		42	106	149	384
Pumping station of treated WW (Cali) viaduct for transfer wastewater in the right side of river	13,419		2,593	2,593	2,593	2,593
Water irrigation network for reuse (Cali, expansion area of Cali and other municipalities)	787		151	151	151	151
Total incremental cost	**19,790**	**5,331**	**2,976**	**3,047**	**3,094**	**3,355**

Table 7.6. Incremental benefits (avoided costs) due to implementation of the Unconventional Strategy 3-SSA (thousands of Euros)

Item	NPV	Year				
		3	5	8	10	20
Initial investment						
DWTP (Cali, expansion area and other municipalities)	6,741	9,408				
-Drinking water distribution network (Cali, expansion area and other municipalities)	3,229	4,507				
-Secondary treatment WWTP-C and WWTP expansion area (Cali)	172,774	241,113				
WWTP for other municipalities	19,298	26,930				
Optimization of WWTP in not-prioritized municipalities	3,134	4,374				
-Wells and pumping stations for irrigation using groundwater (Cali, Cali expansion area and other municipalities)	5,073	6,728		612		
Operation and maintenance O&M						
DWTP (Cali, expansion area and other municipalities)	6,006		1,160	1,160	1,160	1,160
-Drinking water distribution network (Cali, expansion area and other municipalities)	1,166		225	225	225	225
-Secondary treatment WWTP-C and WWTP expansion area (Cali)	69,414		13,411	13,411	13,411	13,411
WWTP for other municipalities	9,216		1,777	1,777	1,777	1,777
Optimization of WWTP in not-prioritized municipalities	1,518		293	293	293	293
-Wells and pumping stations for irrigation using groundwater (Cali, Cali expansion area and other municipalities)	13,745		2,599	2,599	2,694	2,694
Tax for water use in water supply system of municipalities	299		56	57	58	61
Water and sewer tariffs	20,208		3783	3,858	3,909	4,185
Use of fertilizers	22,995		4,439	4,441	4,442	4,449
Payment fee for water use	170		32	33	33	34
Tax for wastewater discharges directly to water bodies	3070		541	573	595	713
Total incremental benefits	**358,056**	**293,060**	**28,316**	**29,039**	**28,597**	**29,002**

7.4 Discussion

The results of this study show that the Unconventional Strategy (3-SSA) has a superior performance compared to the Conventional Strategy with respect to cost effectiveness of treatment and water quality management. Figure 7.3 shows for the Upper Cauca river basin case the factors that have a relatively large impact on this positive result. Among these main factors are the higher initial investment and O&M costs of the WWTPs for the Conventional Strategy compared with the Unconventional Strategy. Moreover, infrastructure of wastewater treatment was much smaller for the Unconventional Strategy. This reduction was due mainly to the joint effects of the prevention/minimization measures (Step 1 of 3-SSA): the change of habits, introduction of low consumption devices, combined with grey water reuse and rainwater harvesting. For the Unconventional Strategy, initial investment and operation and maintenance O&M costs of the WWTPs represent approximately 77% of incremental benefits, of which 68% was associated with Cali city and 9% with other municipalities in the study area. For Step 1 the two factors contributing most to the CBA results are water and sewer tariffs, respectively. They correspond to 6.4% of incremental benefits.

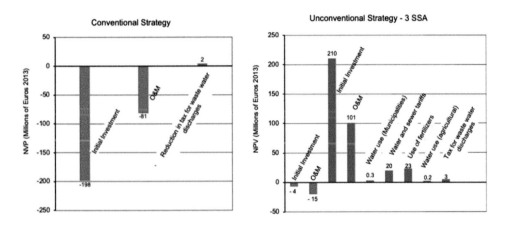

Project horizon: 20 years; social discount rate: 11.75% (1 Euro = 2,500 COP, 2013)

Figure 7.3 Net Present Value (NPV) of incremental cost and benefits for Conventional and Unconventional Strategies

As for Step 2 (treatment for reuse), the most important factor is the avoided cost by use of fertilizer. Avoided costs by taxes for water use and taxes for wastewater discharges directly to water bodies are negligible, since these unit costs are extremely low. For example, taxes for agricultural irrigation are about 300 times lower, as a percentage of minimum wages, compared with raw water prices in Europe and the United States. Despite this, sugarcane farmers report that irrigation represents between 30% and 60% of total costs of cultivation (Cruz, 2015). Due to rapidly growing water demands from municipal, agricultural and industrial uses, and consequent water scarcity, farmers have recently started to introduce efficient irrigation management practices. They are looking to change the irrigation by furrows, with efficiencies of approximately 40% (Diaz, 2006), to sprinkler irrigation systems with efficiencies between 80% and 85% and drip irrigation which can achieve efficiencies of 90%. Colombia is among the countries with the highest use of raw wastewater in agricultural irrigation (Jimenez and Asano, 2008), while irrigation with treated wastewater is virtually non-existent (Universidad del Valle and MADS, 2013). Recently the Government of Colombia introduced new regulations for the use of treated wastewater (MADS, 2014) through which it aims to encourage reuse in both agricultural irrigation and other types of use. The use of treated wastewater must simultaneously ensure that discharge of toxic compounds by households (e.g. metals, chemicals) and industries is substantially reduced, to eliminate potential public health risks.

Step 3 relates to the stimulation of self-purification aimed at speeding up the recovery process of water resources after an organic pollution episode. In this process, organic compounds are diluted and progressively transformed by microbial decomposition. In the rivers, the self-purification capacity depends mainly on: a) the flow, which will dilute the discharged pollution and will facilitate its subsequent degradation to reduce its negative impact on water quality; b) water turbulence, which provides oxygen to the water favouring microbial activity, c) biological activity, in particular from algae and aquatic plants which introduce oxygen into the water column during daytime, d) river morphology (flood plains provide shallow areas with increased capacity for self-purification), and e) the nature and amount and time distribution of the discharges. Although there are limitations to its implementation in the case of the Cauca River, these last two strategies could be the most effective. For the stretch of the Cauca River considered in this study, self-purification capacity was heavily affected by abrupt changes in its dilution ability and by the type, size and spatial distribution of the received pollution. For the Cauca River, most of the self-purification capacity was lost in the last 60 years. For instance, a wetland area of 300 km^2 in the 1950s was reduced in 1986 by 90% (Muñoz, 2012).

In this research the self-purification capacity was associated with the prioritization of investments to maximize impact in improving the water quality of the Cauca River, considering the upper river basin as the unit of analysis. With this approach the interventions in watersheds and municipalities with the highest pollutant load and located upstream of the minimum DO (Puerto Isaacs Station) were prioritized. This strategy arises taking into account the limitations of the Salvajina Dam, located 139 km upstream of the Juanchito Station, to stimulate the self-purification capacity of the Cauca River (Galvis et al., 2014b). The options for re-aeration by turbulence are limited because the Cauca River slope is reduced from $7x10^{-4}$ m/m on the Salvajina-La Balsa stretch to $1.5x10^{-4}$ m/m on the La Balsa-Mediacanoa stretch (Ramírez et al., 2010). This low slope coincides with the stretch where the river receives 70% of its pollution load (Figure 7.1 and Table 7.1)

The strength of 3-SSA (Unconventional Strategy) was the joint and systematic application of the three steps in the context of the basin. In the Cauca River case, the advantages in comparison to the Conventional Strategy are very clear. The unconventional strategy achieved lower BOD$_5$ discharges, higher minimum DO value and a better CBA. The increases of minimum DO (0.4 mg/L) could be equivalent to the additional investment requirement in the 'end of pipe solutions' (WWTPs) in the Conventional Strategy to achieve the same concentration of DO at the critical point, reached with the Unconventional Strategy (3-SSA). This implies that the CBA would be even more favourable for the 3-SSA.

The present study was developed considering only point source pollution and basic parameters such as BOD and DO for dry season conditions. In addition, the water quality modelling was conducted for steady flow conditions in the Cauca River and its tributaries. Further studies will be required to assess the benefits of the 3-SSA considering conditions of unsteady flow conditions and the combined impact of wastewater and urban and rural (agricultural) runoff (Galvis *et al.,* 2014b). Under these conditions, the use of sustainable urban drainage systems (SUDS) as part of the 3-SSA could be considered. Other strategies to consider are: real-time control (automation) of urban drainage and implementation of early warning systems (Velez *et al.*, 2014), and the impact of optimising ecohydrological flows in river-associated wetlands to increase self-purification (Step 3).

For efficient water management in the study basin it is necessary to assign real values to raw water, especially to that used in agriculture. If this decision is implemented, Step 1 (minimization and prevention) and Step 2 (treatment for reuse) will increase their viability (CBA).

7.5 Conclusions

Although actions aimed at pollution control in the Upper Cauca river basin date back over 40 years, the river water quality in the study area continues to decline. This situation persists despite the fact that 19 of the 41 municipalities have WWTPs. In spite of substantial investment in WWTP infrastructure and its O&M costs, the water quality of the Cauca River does not meet the requirements of its uses, including water supply for 76% of the population of Cali city. This approach is limited because it is focused on 'end of pipe solutions' and it does not consider the basin as the unit of analysis and the investments are not executed taking into account their priority and their true impact on the quality of water resources.

The difference between the NPV of incremental benefits and NPV of incremental costs was +338,266x10^3 Euros for the Unconventional Strategy (3-SSA) and -276,318x10^3 Euros for the Conventional Strategy. These results show a clear advantage of the 3-SSA. The CBA mainly reflected the impact of prevention and minimization (Step 1) and the reuse of treated wastewater (Step 2).

Using WWTP effluent for irrigation prevented discharge of residual pollutants into the river, especially upstream of the Paso La Torre Station, and also created economic benefits. The Unconventional Strategy, based on the 3-SSA, resulted in a larger increase of the minimum DO to 2.0 mg/L (Paso de La Torre Station, km 171) for 2033, obtained with the smallest load discharged into the Cauca River. The minimum DO for the Baseline (2013) was 0.6 mg/L (Puerto Isaacs Station, km 155), for Conventional Strategy (2033) it was 1.6 mg/L (Paso de La Torre Station, km 171).

For the Upper Cauca river basin, CBA results also clearly favoured the 3-SSA (Unconventional Strategy). This result is mainly due to the large differences in initial investment and O&M costs of WWTP in municipalities for the two strategies. For the Unconventional Strategy the WWTPs are smaller due the application of the prevention and minimization approaches and treatment for reuse. The impact of the designed treatment system for Cali is very important, considering the population size and costs of activated sludge technology selected for secondary treatment in the Conventional Strategy.

The application of the 3-SSA resulted in avoided costs for initial investments and O&M, especially for groundwater wells and associated pumps for sugar cane irrigation. Furthermore, costs were avoided by optimisation of WWTPs, tariffs and finally by replacement of fertilisers. Avoided costs by taxes for water use and taxes for wastewater discharges directly to water bodies are negligible, since these unit costs were extremely low in Colombia.

The study showed overall positive effects of the 3-SSA on wastewater management in the Cauca basin, primarily through its prevention measures and reuse of the treated wastewater.

7.6 References

Alcaldía de Santiago de Cali (2013). Informe del Departamento Administrativo de Planeación. Cali, Colombia. (In Spanish).

Allen, R.G., Pereira, L., Raes, D. and Smith, M. (2006). Evapotranspiración del cultivo. Guías para la determinación de requerimientos de agua de los cultivos. Estudio FAO Riego y Drenaje, 56. (In Spanish).

Asano, T., Burton, F., Leverenz, H., Tsuchihashi, R. and Tchobanoglous, G. (2007). Water reuse: Issues, technologies and applications. Metcalf and Eddy Inc., AECOM Press and McGraw Hill Professional 1570 p., New York NY, USA. ISBN: 9780071459273.

Ayers R. and Wescot D. (1987). La calidad del agua en la agricultura. Estudio FAO Riego y Drenaje 29.1. Food and Agriculture Organization (FAO), 174. (In Spanish).

Boardman A., Greenberg D., Vining R., Weimer D. (2001). Cost- benefit Analysis: Concepts and Practice (2nd Ed.). Upper Saddle River, N.J., Prentice Hall.

Bos, J.J., Gijzen, H. J., Hilderink, H. B. M., Moussa, M., Niessen, L.W., and de Ruyter-van Steveninck, E.D. (2004). Quick Scan Health Benefits and Costs of Water Supply and Sanitation. Netherlands Environmental Assessment Agency (RIVM), IMTA - UNESCO-IHE, in consultation with WHO, Bilthoven, the Netherlands.

Brent, R. (2006). Applied cost-benefit analysis, Second Edition Ed., Edward Elgar Publishing limited, Cheltenham, UK.

Capra, A. and Scicolone, B. (2007). Recycling of Poor Quality Urban Wastewater by Drip Irrigation Systems. *Journal of Cleaner Production* 15: 1529-1534.

Cardona, M.M. (2007). Minimización de Residuos: una política de gestión ambiental empresarial. *Producción más Limpia* 1(2), pp. 46-57. (In Spanish).

Colpozos (2010) Personal communication about the cost model of pumping systems and irrigation in Cauca valley, Cali. (In Spanish).

Comisión de Regulación de Agua Potable y Saneamiento Básico (2013). Definición de la tasa de descuento aplicable a los servicios públicos domiciliarios de acueducto y alcantarillado. Ministerio de Vivienda, Ciudad y Territorio. Colombia. (In Spanish).

Corporación Autónoma Regional del Valle del Cauca - CVC (2010). Red de monitoreo de la calidad del agua de los recursos hídricos superficiales. Cali. (In Spanish).

Corporación Autónoma Regional del Valle del Cauca - CVC (2012). Acuerdo No. 011 de 2012, Por el cual se fija la tarifa de la tasa por uso del agua. Cali. (In Spanish).

Cruz, J.R. (2015). Manejo eficiente del riego en el cultivo de la caña de azúcar en el valle geográfico del río Cauca. Cali. Centro de Investigaciones de la Caña de Azúcar. Cenicaña. Cali. (In Spanish).

Departamento Nacional de Estadistica DANE (2005). Boletín Censo General. Bogotá, Colombia. (In Spanish).

Díaz, J. (2006). Riego por gravedad, Universidad del Valle. Cali (In Spanish).

Doorenbos, J. and Kassam, A. (1979). Yield response to water, FAO irrigation and drainage. Paper 33. Food and Agriculture Organization FAO, Rome.

EMCALI (1999). Normas para el Diseño de Alcantarillado. Cali. (In Spanish).

Foster, S. and Skinner, A. (1995). Groundwater protection: the science and practice of land surface zoning. IAHS Publications-Series of Proceedings and Reports-Intern Assoc. *Hydrological Sciences*, 225, 471-482.

Galvis A., Zambrano D., van der Steen N. P. and Gijzen H.J. (2014). Evaluation of a pollution prevention approach in the municipal water cycle, *Cleaner Production*. 66,1, 599-609.

Galvis A., Hurtado I. C., Martínez C. A., Urrego, J. G., Van der Steen P. and Gijzen H. J. (2014). Mathematical modelling to study of self-purification capacity of water bodies. Case: Cauca River and Salvajina dam (Colombia). 11th International Conference on Hydroinformatics, August 17-21, 2014, in New York, USA.

Gijzen, H.J. (2001). Anaerobes, aerobes and phototrophs (A winning team for wastewater management), Water Science and Technology, 44 (8) 123-132)

Gijzen, H.J. (2006). The role of natural systems in urban water management in the City of the Future - A 3-Step Strategic Approach. *Ecohydrology and Hydrobiology*, Vol. 6, No. 1-4, pp. 115-122.

González, S., López-Roldán, R. and Cortina, J.L. (2012). Presence and biological effects of emerging contaminants in Llobregat river basin: A review. *Environmental Pollution*, Vol. 161, pp 83-92.

Ifabiyi, I.P. (2008). Self-purification of a freshwater stream in Ile-Ife: lessons for water management. *Human Ecology*, Vol. 24, No. 2, pp.131-137.

Jiménez, B., and Asano, T. (2008) Water Reuse: An International Survey of Current Practice, Issues and Needs, IWA (International WATER ASSN).

Khan, S., and Singh, S.K. (2013). Assessment of the Impacts of Point Load on river Yamuna at Delhi Stretch, by DO-BOD Modelling of river, Using MATLAB Programming. *Engineering and Innovative Technology*, Vol. 2, No. 10, pp. 282-290.

Liu, S., Butler, D., Memon, F. A., Makropoulos, C., Avery, L. and Jefferson, B. (2010). Impacts of residence time during storage on potential of water saving for grey water recycling system. *Water Research*, 44(1), pp. 267-277

Mejia, F.J., Isaza, P.A., Aguirre, S. and Saldarriaga, C.A. (2004). Reutilización de aguas domésticas. In: XVI Seminario Nacional de Hidráulica e Hidrología, Armenia, Colombia. (In Spanish).

Ministerio de Ambiente y Desarrollo Sustentable MADS (2014). Resolución N° 1207 July 25, 2014. Por la cual se adoptan disposiciones relacionadas con el uso de aguas residuales tratadas. Bogota, Colombia. (In Spanish).

Ministerio de Agricultura y Desarrollo Rural (2010). Evaluación Departamental de Costos de Producción de Caña de Azucar. Cali, Colombia. (In Spanish).

Muñoz, E. (2012) Wetlands in the American continent. Concepts and classification in Peña, E.J., Cantera, J.R., Muñoz, E. (Eds.) Evaluation of pollution in aquatic ecosystems. Case study in the Sonso Lagoon, Upper Cauca river basin (pp. 23-49). Universidad Autonoma de Occidente and Universidad del Valle, Editorial Program, Cali, Colombia. (In Spanish).

Nhapi, I. and Gijzen, H.J. (2005). A 3-Step Strategic Approach to sustainable wastewater management. *Water SA*, 31(1), pp. 133-140.

Ostroumov, S.A. (2008). Basics of the Molecular-Ecological Mechanism of Water Quality Formation and Water Self-Purification, *Contemporary Problems of Ecology*, Vol. 1, No. 1, pp. 147–152.

Ottoson, J. and Stenström, T.A. (2003). Faecal contamination of grey water and associated microbial risks. *Water Research*, 37(3), pp. 645-655.

Ramirez, C.A., Santacruz S., Bocanegra, R.A. and Sandoval, M.C. (2010). Salvajina reservoir incidence on the flow regime of the Cauca river in its upper valley. *Ingeniería de los Recursos Naturales y del Ambiente*, Vol. 9, pp. 89-99. (In Spanish).

Sánchez, F. (2013) Detailed costs for some WWTPs designed and built by Sánchez Engineering in Valle del Cauca. Personal consultation held in August 2013.

Sandoval, M.C. and Ramírez, C. (2007). El río Cauca en su valle alto: Un aporte al conocimiento, Cali. (In Spanish).

Sandoval, M.C., Ramirez, C.A., and Santacruz S. (2007). Monthly operating rule optimization of Salvajina reservoir. *Ingeniería de los Recursos Naturales y del Ambiente*, Vol. 6 (2007) pp. 93-104. (In Spanish).

Sokolov, A. and Chapman, T. (1981). Métodos de cálculo del balance hídrico. Guía internacional de métodos de investigación. Instituto de Hidrología de España-UNESCO. (In Spanish).

Torres, J. and Carbonel, J. (1996). Avances técnicos para la programación y manejo del riego en caña de azúcar. Serie Técnica-Centro de Investigación de la Caña de Azúcar de Colombia. (In Spanish).

Universidad del Valle and Corporación Autonama Regional del Valle del Cauca CVC (2007). Optimization of the water quality simulation model of the Cauca river. La Balsa – Anacaro reach. Project report Volume XIII. Cauca river Modelling Project (PMC), Phase III. Santiago de Cali, Colombia. (In Spanish).

Universidad del Valle and Corporación Autonama Regional del Valle del Cauca CVC (2009). Water quality modelling scenarios of the Cauca river. Project report. Cali, Colombia. (In Spanish).

Universidad del Valle and EMCALI (2006). Impact assessment of the proposed strategies by EMCALI for the management of wastewaters in the city of Cali on the water quality of the Cauca River. Project report. Cali, Colombia. (In Spanish).

Universidad del Valle and Ministerio de Ambiente y Desarrollo Sostenible MADS (2013). Evaluación del estado de los procesos de reuso de aguas residuales en Colombia. Project Report. Cali, Colombia. (In Spanish).

U.S. EPA (1998). Water Recycling and Reuse: The Environmental Benefits. Report produced by Environmental Protection Agency. Washington D. C., USA.

U.S. EPA (2012). Guidelines for Water Reuse. U.S. Environmental Protection Agency. Washington, USA.

Velez C.A., Alfonso L., Sanchez A., Galvis A. and Sepulveda G. (2014). Centinela: an early warning system for the water quality of the Cauca River. *Journal of Hydroinformatics*. 6, 1409-1424.

Von Sperling, M. (2005). Introduction to water quality and sewage treatment, 3th edition, Department of Sanitary and Environmental Engineering – Federal University of Mina Gerais, Belo Horizonte. (in Portuguese).

WHO (2006). Guidelines for the safe use of wastewater, excreta and grey water. Volume 2. Wastewater use in agriculture. World Health Organization (WHO). Paris, France.

Chapter 8
Conclusions and recommendations

Source: CVC photo file

8.1 Conclusions

The protection of water resources from deterioration in quality by point and non-point source pollution discharges is probably the biggest challenge in sustainable water resources management and it has been growing during recent decades. In practice, most countries have adopted pollution control strategies and measures which are based on 'end-of-pipe' solutions and consider only wastewater treatment plants (WWTPs). The introduction of end-of-pipe treatment technology is usually accompanied by adjustments to the regulations, including the application of economic instruments, such as taxes for wastewater discharges. However, this strategy has limitations. On many occasions, the end-of-pipe approaches are not able to comply with the proposed objectives because some systems have been abandoned and others are operating with lower efficiency than for which they were designed. Another issue is the very high costs and lack of rational prioritization of investments.

The research described in this thesis was designed to contribute to the development of sustainable solutions for the previously outlined problem. Therefore, it was oriented towards the development of a strategy of interventions and technology selection based on the Three-Step Strategic Approach concept (3-SSA), which is not only focused on the urban water cycle, but also on the basin, considering it as the unit of analysis. The 3-SSA includes: 1) prevention or minimisation of waste production; 2) treatment aimed at recovery and reuse of waste components, and 3) disposal of remaining waste with stimulation of natural self-purification of the receiving water body. These three steps should be implemented in the above sequence, and possible interventions under each step should be fully exhausted before moving on to the next step. However, in this research each step was studied first independently (chapters 3, 4, 5 and 6), before making a full assessment of the potential impact of the three steps combined (Chapter 7). The results and conclusions of the study of each step were an input to perform a comprehensive analysis of the sequential implementation of the three steps combined.

The research included the identification, application and validation of the 3-SSA by applying it to a specific basin in Colombia. This included, among other factors, the identification of priority water uses and the wastewater pollution control plans for both medium and long term. In this context, 'technology selection' will not be limited to treatment technology but will also include aspects such as minimization and prevention, both in the urban water cycle (housing and urban drainage system) and at the basin level, WWTPs, reuse of effluents, and the natural and/or stimulated self-purification capacity of the water bodies. The study area was the Upper Cauca river basin of the Cauca River, the second most important river in Colombia. The study stretch has a length of 389 km, and is located between the stations La Balsa and Anacaro.

8.1.1 Step 1 - Minimization and prevention: Strategies at the household level (Chapter 3)

For the first step of the 3-SSA, in the case of households, minimization and prevention can be achieved through combinations of the introduction of low consumption devices, use of grey water, and rainwater harvesting. The best alternatives were selected applying the analytical hierarchy process (AHP) and grey relational analysis (GRA), considering multiple criteria including technical, social, environmental and economic. Additionally, a Cost-Benefit Analysis (CBA) was used to evaluate the best minimization and prevention strategies versus the conventional approach (without the use of low consumption devices or use of drinking water for all uses) to determine the viability of this first step. The results demonstrated that generally prevention and minimization measures have advantages compared to the conventional approach in terms of the cost to benefit ratio.

The case study took place in the expansion area of Cali (Colombia), which will have 410,380 inhabitants. It was considered that 70% of households would apply systems that included minimization and prevention alternatives. The evaluation of minimization and prevention alternatives was done using two different system boundaries: (Scenario 1) reduction in water supply costs for households and the avoided costs in the infrastructure of additional sewerage and wastewater treatment facilities; and (Scenario 2) only taking into account the reduction in water supply costs for households and the savings associated with the drinking water infrastructure.

For AHP + GRA the main results were similar for Scenario 1 and Scenario 2. The main findings were as follows:

-   According to the AHP and GRA processes, the best minimization alternative for Cali's expansion zone corresponds to Alternative **C** (WC dual flush; grey water and rainwater harvesting). Alternative **C** was the best solution when comparing the CBA of the **conventional approach** (toilet 6 L and drinking water for all uses) and the minimization and prevention alternatives **B** (WC dual flush; grey water), **C** (WC dual flush; grey water; rainwater harvesting) **G** (WC 2.3 L; grey water), and **H** (WC 2.3 L; grey water; rainwater harvesting). This was because high efficiency WC equipment (2.3 L) is relatively expensive on the local market.

-   Minimization and prevention alternatives **B, C, G** and **H** are the best alternatives according to the AHP and GRA results, independent of the type of percentage distribution of single and multi-family households, and the percentage of households implementing minimization and prevention alternatives.

For CBA (incremental situation) the main results were as follows:

- Costs for the implementation of minimization and prevention alternatives are associated with the additional internal network infrastructure, including initial investment and operation and maintenance costs. For Alternative B, initial investments are: €439 for single households (low-consumption power device: €44; grey water system: €395) and €291 for the multiple households (low-consumption device: €44; rainwater harvesting: €86; grey water system: €161). The Net Present Values (thousands of Euros) for this alternative corresponded to: Initial investment internal network of water & sanitation = 11,913 x10^3 Euros; O&M= 18,515x10^3 Euros; Replacement: 1,301 x10^3 Euros; Total = 31,729 x10^3 Euros. Then, in relation to the implementation of Alternative B, the most relevant costs (inside the home) were those associated with the operation and maintenance, followed by the initial investment costs.

- The alternatives with the highest ranking in the application of AHP and GRA (**B, C, G and H**) compete in terms of CBA (best cost-benefit ratio) with the conventional approach, when considering the reduction in water supply costs for households and savings in the water supply, sewage and WWTP infrastructure. The ratio NPV $_{Benefit}$/ NPV$_{cost}$ for each alternative of Scenario 1 were: B=1.14; C=1.22; G= 1.09 and H=1.08. For each alternative of Scenario 2 the ratio was: B=1.07; C=1.15; G= 1.03 and H=1.02. All these alternatives were feasible because NPV$_{Benefit}$/ NPV$_{cost}$>1 for all cases. For the two scenarios the best alternatives were B and C, in this order.

- Minimization and prevention alternatives become viable when the percentage of multi-family households using such alternatives increases. For the study area, the minimization and prevention alternatives were viable (NPV Benefit/NPV cost >1.0) when these are implemented in more than 20% of households.

- In urban models with a greater number of single-family households, the most feasible alternative is **B** (WC dual flush; grey water), while in the event of a larger number of multi-family households, the best alternative is **C** (WC dual flush; grey water and rainwater harvesting). The urban models generating the greatest benefits are those corresponding to 100% single-family homes (alternatives **B, G, H**) and to 100% multi-family homes (Alternative **C**). In general, with minimization and prevention approaches, the water demand decreases according to the percentage of households that implement it. The urban model with the highest percentage of multi-family dwellings generates the lowest wastewater volume per capita. In this type of urban development, grey water is used for irrigation purposes and cleaning of communal areas, while single-family dwellings generate grey water in excess, due to the absence of communal areas in this type of household.

8.1.2 Step 1 - Minimization and prevention: Strategies at urban drainage level (Chapter 4)

Technology selection for urban drainage systems plays an important role in the efficient management of runoff and wastewater. It is a complex decision involving different criteria, including environmental, social, technical, economic and institutional. This process should include several technological options and consider the interaction between the sewer, the WWTP and the receiving water body. Multi-criteria methodology allowed the use of scientific knowledge and experience of local experts in the design and construction of a conceptual framework (CF) for technology selection. This CF was based on the Three-Step Strategic Approach (3-SSA) and can be applied in new urban areas and in expansion areas of existing cities. The criteria and the information requirement are easily recognized by both decision-makers and designers in Colombia and the Latin American context.

The CF for technology selection of urban drainage considers the following sequence:

Block 1. Pollution prevention and waste minimization at different levels. The CF considers: erosion control and watershed maintenance; comprehensive management of solid waste; cleaning of roads; management of household chemicals and efficient use of water;

Block 2. SUDS selection. To control water quantity and to improve the water quality of urban runoff through infiltration and storage devices, the users of the CF may choose one or more of the following SUDS options: permeable paving, infiltration tanks, detention tanks, retention ponds, and constructed wetlands;

Block 3. Assessing of the surface drainage feasibility. In this block the CF evaluates the possibility of using the hydraulic capacity of roads and ditches to drain (fully or partially) the surface flow that has not been captured by SUDS;

Block 4. The choice between combined or separate sewer systems. Runoff that was not managed by SUDS or surface drainage (blocks 2 and 3) has to be collected and transported by a sewer system to a final disposal point. Among the different indicators used in this block, two are directly connected with pollution control: i) First flush control and ii) Dilution and self-purification capacity of the receiving water body;

Block 5. Selection of the type of sanitary sewer. For wastewater management the options considered were combined sewers and separate sewers (septic tanks and small diameter pipes; simplified sewers, conventional sewers).

The case study was 'Las Vegas' (59 ha; 15,000 inhabitants), an expansion area of Cali, Colombia. The following was concluded:

- The results obtained with the CF for each block were as follows: erosion control and watershed maintenance (Block 1); SUDS (detention tank) to handle 37% of runoff in the drainage area (Block 2) and combined sewer (blocks 4 and 5). For Block 4, The main indicators for the technology selection were:

    Dominant slope of drainage area: 0.003-0.005 m/m
    Storm sewer pipe diameter: 600-800 mm
    Wastewater pumping requirements: 0%
    Drainage area managed by SUDS: 37%
    SUDS selected in Block 2. Detention tank
    Receiving water body discharge during the dry season: 0.513 m3/s
    Maximum flow of runoff: 1.5 m3/s
    Ability to control illegal connection to sanitary sewer system: Medium

    The weighted sum score of the alternatives considered in this block ($P_{i\,WSM}$) were for **combined sewerage (CS)**: Topography: 95; Wastewater pumping requirements: 62; First flush control: 46; Dilution capacity and self-purification capacity of receiving water body: 34; O&M complexity: 37; Ability to control illegal connections to sanitary sewer: 92.5. For **Separate System (SS)**: Topography: 95; Wastewater pumping requirements: 46.5; First flush control: 34.5; Dilution capacity and self-purification capacity of receiving water body: 51; O&M complexity: 55.5; Ability to control illegal connections to sanitary sewer: 55.5. In summary (*Pi WSM*) for **CS: 366.5** and for **SS: 338** According to these results the technology selected in Block 4 was the combined sewerage.

- The application of the CF in Colombia and other Latin American countries can contribute to: 1) improving urban drainage system planning, 2) considering the broader technological offer, beyond the traditional one (separate sewer and combined sewer), selection of the type of sanitary sewer also includes: small diameter sewers with interceptor tanks and simplified sewer; 3) improving the selection of urban drainage system technology.

8.1.3 Step 2 - treatment, recovery and reuse of waste components (Chapter 5)

The potential for reuse of treated wastewater can be assessed through the agricultural plan and the potential area of irrigation obtained from the mapping of five parameters: 1) Current land use; 2) Proximity of the point of delivery of treated wastewater for irrigation 3) Land slope in the direction to the area when the treated wastewater will be used for irrigation; 4) Physical limits for the use of treated wastewater for irrigation; 5) Vulnerability to contamination of the aquifer system. With the irrigation requirements and the value of the area to be irrigated, the flow was obtained. The reuse potential was complemented by a CBA of the incremental situation when comparing the options with and without reuse of treated wastewater.

Water requirements for irrigation is a key factor in reuse viability. Also the values of economic instruments (water tariffs and taxes for wastewater discharges to water bodies) can affect the CBA results, and with this the reuse feasibility of treated wastewater. For example, raw water tariffs for agricultural irrigation in Colombia are about 300 times lower, as a percentage of minimum wages, than those in Europe and the United States. With this, the efficient use of water is not stimulated and, in some cases, the reuse of effluents for irrigation is not feasible.

The application of this methodology to evaluate the reuse potential in irrigation of sugarcane crops in three case studies (1. Cali; 2. Expansion area of Cali; and 3. Buga City) in the Upper Cauca river basin, in Colombia, showed the following results:

-   Cost Benefit Analysis (CBA): the differences between $NPV_{benefits}$ and NPV_{cost} were: Case 1: -20,474,344 Euros; Case 2: - 407,037 Euros and Case 3: 1,437,740 Euros. For the ratio $NPV_{benefits}/NPV_{cost}$ the results were: Case 1= 0.60; Case 2= 0.82 and Case 3= 1.50. These results show the economic feasible for Case study 3 ($NPV_{benefits}/NPV_{cost}$> 1.0) and the economic infeasibility for the case studies 1 and 2 ($NPV_{benefits}/NPV_{cost}$ < 1.0).

-   Cases 1 and 2 have very different irrigation requirements: 1.08 L/s-ha and 0.34 L/s-ha, respectively, although they are located in the same geographical region. This suggests that the results obtained in a specific location cannot be generalized to other locations.

-   The sensitivity analysis regarding costs of water tariffs and taxes for wastewater discharges to water bodies shows a strong impact of these fees (tariffs and taxes) on CBA. The fees during the reference year (2013) for this study were extremely low, which did not favour the viability of irrigating sugarcane crops in the Upper Cauca river basin. Case 1 would be feasible ($NPV_{benefits}$ / NPV_{cost}> 1.0) if the value of water tariffs increases by approximately 75 times the reference value (year 2013). Case 2 would be feasible if the value of water tariffs also increases by approximately 75 times the reference value (year 2013) or if the value of tax for wastewater discharges increases by approximately 40 times the reference value (year 2013). Current water and effluent discharge tarifs/taxes in Colombia are about 300 fold lower than those in the USA or EU, as a percentage of minimum wages. This shows that this could be reasonably increased 75 times the current value.

8.1.4 Step 3 - Disposal of waste with stimulation of natural self-purification (Chapter 6)

The case of the Upper Cauca river basin (in Colombia) was studied using dynamic modelling. In this basin, where approximately half of the municipalities already have a WWTP, water quality is now worse, compared to the time when these treatment systems did not exist. This deterioration can be explained by the pollution load increase (domestic and industrial) generated in the basin and the limited effectiveness of wastewater treatment plants, including the one in the city of Cali (WWTP-C). However, this deterioration is also associated with variations in the Cauca River flow due to the upstream Salvajina reservoir and the impact of its tributaries, in particular the South Drainage System of the city of Cali. Below are the baseline conditions of the study.

*Base line steady-state condition (*permanent flow)
Pre-Salvajina condition (before construction of the dam, 1985) corresponding to flow in Juanchito Station (km 139): 88 m^3/s, and Post-Salvajina condition (after construction of the dam, in 1985), flow in Juanchito Station (km 139): 143 m^3/s).

Base line dynamic condition (non-permanent flow)
Post-Salvajina condition (after construction of the dam, in 1985), mean flow of 143 m^3/s at Juanchito station (km 139) and considering flow non-permanent and variations in water temperature, BOD$_5$ and DO in the upper boundary condition.

Step 3 of the 3-SSA was studied for the Upper Cauca river basin, in the Balsa - Anacaro stretch (389 km). The research was focused on comparison of the results of dynamic modelling and steady state modelling to study the impact of pollution on the river. The research included the impact of the Salvajina dam on the water quality of the Cauca River. The main conclusions were as follows:

- The Cauca River has a dynamic behaviour associated with the operation of the Salvajina reservoir, located 27.4 km upstream from the La Balsa Station and 139 km upstream from the Juanchito Station. This latter station was used as a reference station because it is located at a short distance (a few hundred meters) from the intake of city of Cali water supply system. In addition, the Cauca River receives pollutant discharges with dynamic behaviour, even during the dry season, including typical variations of discharges from WWTP-C, including pollutant peaks due to out-of-operation periods of WWTP-C. However, one of the most critical dynamic situations is the pollutant flush happening during rainfall events (first flush effect), associated with diffuse and accumulated pollution in the urban and rural sectors of the South Drainage System of Cali.

- The dynamic behaviour of water bodies and the pollution sources significantly affect the self-purification capacity of the water bodies (Step 3 of 3-SSA). Flow changes and pollution peaks generate variations in the dilution capacity and DO levels. When the pollution peaks coincide with periods of low flow, the minimum DO and self-purification capacity (Step 3 of 3-SSA) in the Cauca River was reduced and anoxic water conditions reached the upstream point of water intake for Cali.

- The pulsed regime effect of the Salvajina reservoir on the hydraulic behaviour of the Cauca River and its impact on the water quality and self-purification capacity must be studied

- under dynamic conditions. The results of the water quality modelling are indicated below. For the assumption of steady-state condition (flow of 143 m^3/s at Juanchito station, km 139) the Cauca River would remain under aerobic conditions (DO>0) in the study stretch (La

  Balsa – Anacaro). However, for the assumption of dynamic condition (average flow of 143 m^3/s at Juanchito station, km 139) DO value close to zero is expected at Puerto Isaacs Station (km 155).

- The pollution associated with rainfall events (first flush effect) in the southern drainage system of the city of Cali (South Channel), scenarios S3 and S4, generated abrupt reductions in DO concentrations at Juanchito Station (corresponding with the suspension of the water intake of the Cali water supply system). Values below 1.0 mg/L DO were found at this point, which results in frequent closures of the potable water purification plant. The pollution impact due to rainfall events is less critical when the peak of pollution is generated during daytime, when the Cauca river flows are higher, by the way of operation of the Salvajina reservoir. This condition also shows the effect on the water quality of the Cauca River as a source of water supply for the city of Cali due to the reservoir operation.

- Step 3 of the 3-SSA can play an important role in cost-effective water quality management of rivers and associated water bodies. To study this, we must use dynamic modelling as a tool and consider the basin as a unit of analysis.

8.1.5 Sequential implementing of the three steps (Step 1 + Step 2 + Step 3)

The strength of 3-SSA is based on the sequential implementation of the three steps. Also, possible interventions under each step should be fully exhausted before moving on to the next step. In this research the 3-SSA (non-conventional strategy) was validated by applying it to the Upper Cauca river basin (La Balsa – Anacaro Stretch) in Colombia and comparing it to a Conventional Strategy, which considers a 'business as usual scenario' of high water use, 'end-of-pipe' wastewater treatment and conventional water supply providing drinking water quality for all uses. In this research the 3-SSA is validated as an Unconventional Strategy, which includes reduction in water consumption (Step1) and reuse of treated wastewater in households and for sugarcane crop irrigation (Step 2). It also considers prioritization of investments to maximize impact in improving the water quality of the Cauca River in the study area, targeting interventions in watersheds and municipalities with the highest pollutant load and located upstream of the river segments with the lowest DO (Step 3). The following are the main conclusions:

- Although actions aimed at pollution control in the Upper Cauca river basin date back over 40 years, the river water quality in the study area is continuing to decline. This situation persists despite the fact that 19 of the 41 municipalities have WWTPs. In spite of substantial investment in WWTP infrastructure and its O&M costs, the water quality of the Cauca River does not meet the requirements for its uses, including water supply for 76% of the population of Cali (2 million aprox.). This approach has failed not only because it is focused on 'end of pipe solutions' but also because it does not consider the basin as the unit of analysis and the investments are not executed taking into account their priority and their eventual impact on the quality of water resources.

- The pollutant load removed was 64,805 kg/d BOD_5 (46%) for the Conventional Strategy and 69,402 kg/d BOD_5 (50%) for Unconventional Strategy. Cost benefit analysis results clearly favoured the 3-SSA (Unconventional Strategy): NPV for Conventional Strategy $=-276,318\times10^3$ Euros, and NPV for Unconventional Strategy (3-SSA) $=+338,266\times10^3$ Euros. The application of the 3-SSA resulted in avoided costs for initial investments and O&M, especially for groundwater wells and associated pumps for sugarcane irrigation. Furthermore, costs were avoided by optimization of WWTPs, tariffs and by replacement of fertilizers.

- The modelling result corresponding to baseline Post-Salvajina condition (year 2013) showed a minimum DO 0.6 mg/L in Puerto Isaac Station (km 155). The modelling results corresponding to the Unconventional Strategy (3-SSA) showed a minimum DO 2.0 mg/L in Paso de La Torre Station (km 171) for the year 2033, while the modelling results corresponding to the Conventional Strategy showed a minimum DO 1.6 mg/L in Puerto Isaacs Station (km 155) for the year 2033.

- The results of this study show that the Unconventional Strategy (3-SSA) has a superior performance compared to the Conventional Strategy with respect to cost effectiveness of treatment and water quality management. There are some factors that have a relatively large impact on this positive result. Among these main factors are the higher initial investment and O&M costs of WWTPs for the Conventional Strategy compared with the Unconventional Strategy. Moreover, infrastructure of wastewater treatment was much smaller for the Unconventional Strategy. This reduction was due mainly to the joint effects of the prevention/minimization measures (Step 1 of 3-SSA): introduction of low consumption devices, combined with grey water reuse and rainwater harvesting.

- Regarding the CBA of the incremental situation, for the Unconventional Strategy (NPV=+338,266$\times10^3$ Euros), initial investment and O&M costs of the WWTPs represent approximately 77% of incremental benefits, of which 68% was associated with Cali and 9% with other municipalities in the study area. For Step 1 the two factors contributing most to the CBA results are water and sewer tariffs, respectively. They correspond to 6.4% of the incremental benefits.

- The impact of the designed treatment system for Cali is very important, considering the population size and costs of activated sludge technology selected for secondary treatment in the Conventional Strategy.

- The application of the 3-SSA resulted in avoided costs for initial investments and O&M, especially for groundwater wells and associated pumps for sugarcane irrigation. Furthermore, costs were avoided by optimisation of WWTPs, by tariffs and by replacement of fertilisers. Avoided costs by taxes for water use and taxes for wastewater discharges directly to water bodies are negligible, since these unit costs are extremely low in Colombia.

- In countries such as Colombia, the results of this research can contribute to strengthening the formulation and application of solid policy and management tools so that government agencies and environmental authorities are oriented towards strategies proposed in the 3-SSA.

8.1.6 Final considerations

The results of this research and the proposed methodology intend to contribute to the development of strategies that optimize the investments related to water resources management. This is a way to contribute to the challenges of the Sustainable Development Goals (SDGs). These SDGs, with a projection until 2030, recognize the centrality of water resources for sustainable development and the vital role that improved drinking water, sanitation and hygiene play in progress in other areas, including health, education and poverty reduction.

In Colombia and other Latin American countries, most institutions include in their public policy and regulations concepts such as: Resilience, Integrated Water Resources Management (IWRM), Hydrological Cycle, Urban Water Cycle (UWC), Integrated Water Management Urban Water (IWM), Ecohydrology, Governance, etc. However, usually these are not captured in an overarching strategy, and often these do not translate into concrete actions. The methodology and results of this research provide an opportunity to put some of these concepts into practice as part of an overarching strategy (the 3-SSA).

The use of the results of this research can contribute to the challenge of the paradigm shift towards sustainable cities, by developing comprehensive plans at river-basin level. This change of paradigm and the application of the 3-SSA must consider the basin as a unit of analysis, inter-institutional and interdisciplinary work that makes it possible to reach a shared vision and to act with transparency. In doing so, the quality of the water, sanitation and other eco-systemic services of the water resource can be favourably impacted, avoiding inequities, exclusion and minimizing externalities. For Colombia and other Latin American countries, this represents a great challenge. To face this challenge effectively, these countries must make many changes. Among the current conditions that must be changed are the following: 1) water management is carried out in a fragmented manner by different sectors according to the type of use (domestic, agricultural, power generation, etc.); 2) environmental authorities operate by political administrative divisions. The basin is not the unit of analysis; 3) there is a lack of leadership and limitations in the training processes of the personnel linked to the institutions of the water sector; 4) generally the vision is only short-term, only covering one government period (3-4 years); 5) the experience of teamwork is limited as well as the effectiveness in the community participation, which makes it difficult to build a long-term shared vision to build a comprehensive solution to the problems of water resource management; 6) there are many management plans but these are made by different sectors, with different purposes, different scales (city, department, basin, etc.) which limits their effectiveness; 7) in the last few decades, investments not related to infrastructure have been reduced, such as: institutional strengthening, research and education, and 8) approaches and interventions are usually not based on a holistic and integrated strategic plan, such as the 3-SSA.

8.2 Recommendations

8.2.1 Application of the obtained results

The investments to improve the water quality for their different uses have focused largely on 'end of pipe' approaches via the construction of WWTP. The results of the research presented in this thesis suggest that 'technology selection' should not be limited to wastewater treatment technology only, but it must include aspects such as minimization and prevention, both in the urban water cycle (housing and urban drainage system levels) and in the basin context, WWTPs, reuse options of effluents, and the natural and / or stimulated self-purification capacity of the water bodies. The application and validation of the 3-SSA and the comparison with conventional approaches considered, among other factors, the priority water uses and the wastewater pollution control plans involving activities for short, medium and long terms.

To take advantage of the benefits of the proposed strategy in the current research, the basin must be used as a unit of analysis. Additionally, environmental, social, cultural, economic, policy and regulatory aspects should be included. It is necessary for stakeholders to build a shared vision, to encourage community participation and teamwork and to act in a transparent manner. Aspects such as the solid waste disposal to water bodies and the urban drainage system are associated with cultural behaviour and solutions must be found that take this into account. The inclusion of multiple criteria decision analysis (MCDA) can facilitate the participation of different stakeholders in making decisions.

8.2.2 Recommendations for further research

The cost - benefit analysis (CBA). To improve the CBA of the incremental situation, the following action is recommended:

- Consider some benefits not included in this research, such as the favourable impact of eco-system services as such: nutrient recycling, habitat for plant and animal species; food production, recreation, eco-tourism, etc.

- Compile and organize information on the most relevant components involved in the CBA such as low consumption devices, grey water reuse infrastructure, infrastructure for rainwater harvesting, wastewater treatment (with and without reuse), irrigation networks for reuse, etc.

- Develop investment and O&M cost models, given the differences in unit costs in the same region or country.

- Analyse scenarios that show the impact of implementing Step 1 in steps 2 and 3, and the impact of implementing both steps 1 and 2 in the implementation of Step 3. It is expected that this will reinforce the importance of implementing the three steps in sequence, instead of implementing each step in isolation.

Water quality indicators. In this study dissolved oxygen (DO) and biochemical oxygen demand (BOD_5) are presented as classic indicators of water pollution. The DO concentration is a primary measure of a stream's health, and it responds to the BOD_5 load. However, it is recommended for future research to also include other compounds such as pesticides, fertilizers, heavy metals, micro pollutants, etc., which may have other (eco-toxicological) impacts. For these contaminants the best management options are provided under Step 1 of the 3-SSA (minimisation and prevention).

Instrumentation Control and Automation (ICA). It is recommended to improve the instrumentation to allow higher quantity and quality monitoring of the different components of the system. The aim is to get more accurate information on the behaviour of precipitation (urban and rural), the main river and tributaries, water extractions, sewer systems, treatment plants, etc.

Better information will facilitate the dynamic modelling (quantity and quality), allowing a better understanding of the behaviour of the system in terms of: 1) effect of operation of reservoirs; 2) effects in the quantity and quality of the river by variations in its tributaries; 3) first flush effect due to diffuse pollution in urban environments; 4) suspension of sediment material in the urban drainage system; 5) impact in the self-purification capability, from the optimization of the interaction between the river and its wetlands and flood plains; 6) climate change and climate variety scenarios; 7) impacts of sustainable urban drainage systems (SUDS) in the reduction of quantity and quality peaks.

Better information will facilitate the implementation of modelling that integrates the urban drainage system, the WWTP and the water body. Later it is recommended to implement real-time control (automation) as a strategy for water pollution control.

Policies and regulations. The following actions are recommended: 1) review the water tariffs and taxes for water discharges with a view to stimulating the efficient use of water (Step 1) and the reuse of treated wastewater both in agriculture and in other uses (Step 2); 2) stimulate reforestation of the basins related to the urban drainage system (Step 1) and the implementation of practices related to WSUD, including the implementation of SUDS (Step 1); 4) stop the loss of the wetlands and floodplains of the rivers and, where possible, recover the areas that have been lost in the past (Step 3); 5) improve information on the self-purification capacity of the water bodies to consider this capacity (natural and / or stimulated) in the definition of quality objectives, treatment objectives and in the implementation of plans to achieve these objectives (Step 3).

8.2.3 Recommendations for continuity of the case studies in the Upper Cauca river basin

Step 1. Prevention and minimization at household level: 1) Update the selection process by reviewing the costs of low-consumption devices; 2) Consider the implementation of cleaner production in the industrial sector with greater impact to its wastewater discharges.

Step 1. Prevention and minimization at the urban drainage level: 1) develop software to facilitate the application and validation of the Conceptual Framework (CF); 2) develop a validation of the conceptual framework CF to select urban drainage system technology, based on 3-SSA. This will be including its application for new case studies; 3) evaluate the impact of urban diffuse pollution and the impact of the first flush; 4) include other options of SUDS, considering in the current version of the CF only four options of SUS are considered: infiltration tank, detention tank, contrition wetland and permeable paving.

Step 1. Prevention and minimization considering agricultural irrigation water use. Evaluate the impact of improving irrigation efficiency in sugarcane crops. Currently it is done by furrows, with efficiencies of about 40%. This means that every 100 L that are used by irrigation of crops, only 40 L are actually used by the crop, while sprinkler or drip irrigation could raise efficiencies to 60 - 90%.

Step 2. Wastewater treatment for reuse: 1) develop cost models for technology selection of wastewater treatment considering reuse in agricultural irrigation; 2) develop cost models of the technologies with the greatest potential to be implemented in the study area; 3) include additional benefits such as the use of bio-solids, biogas and energy; 4) evaluate the impact of recently updated Colombian policies and regulations, including the first regulations about reuse of treated wastewater of 2014.

Step 3. Stimulated natural self-purification; For the Cauca River, most of the self-purification capacity has been lost in the last 60 years. In this context it is recommended to continue studying other scenarios for the application of ecohydrology concepts, which may help to stimulate natural self-purification. For the study area in this research one could consider, for example 1) optimisation of the ecohydrology of the Sonso Lagoon, the most important wetland in the study area, and *2)* the effect of recovery of floodplains along the Cauca River in the study area.

Curriculum Vitae

Alberto Galvis Castaño

Alberto Galvis Castaño was born on March 31, 1955 in Sevilla, Department of 'Valle del Cauca', Colombia. He grew up in Cali and graduated as a Bachelor in Sanitary Engineering in 1979 and a Master of Science in System and Industrial Engineering in 1988 from the Universidad del Valle. Between 1979 and 1983 he worked in engineering consultancy companies. In the period 1983-1985 he participated as a staff member of an international consortium in the project 'Cali wastewater treatment study'. Alberto Galvis has been associated with the Faculty of Engineering at the Universidad del Valle since 1986.

He was a member of the founding group of the Cinara Institute (1989) and a member of the teams that led the creation of the research groups in 'Water Supply' (1989) and 'Integrated Water Resources Management' (2003). He was a member of the working groups in the Universidad del Valle that led to the creation of Master programs in: 'Sanitary and Environmental Engineering' (1993) and 'Integrated Water Resources Management' (2015). He has been a director of international cooperation programs with institutions in Europe and Latin America and a director of research and development projects with the participation of Colombian institutions in the water resources management sector. He was made a full professor (2001) and Distinguished Professor (2010) of the Universidad del Valle.

He is currently Director of the research group in 'Integrated Water Resource Management' of Cinara Institute, as well as Senior Researcher in the national science and technology system (Colciencias). Alberto Galvis is a lecturer in undergraduate and graduate academic programs and researcher in the topics of technology selection in water and sanitation, and mathematical modelling applied to planning and water resources management. He has supervised more than 35 Bachelor and Master's theses, including co-directing of students from Australia, Belgium and the Netherlands. He was a director of advisory projects to the Ministry of the Environment in Colombia (2011-2014) to review and update the policy and regulations of water resources management, in the following topics: i) water quality for different types of use; ii) reuse of wastewater and iii) protection of groundwater by wastewater discharge into the ground.

The professional performance of Alberto Galvis has been closely related to the Upper Cauca river basin, from his consulting activities, his Master's thesis ('Water quality modelling of the Cauca River'), to his activities in research and development projects, as a staff member of the Universidad del Valle, through water companies, environmental authorities, the Colombian Agency for Science, Technology and Innovation (Colciencias) and international cooperation projects. Among the latter was the SWITCH Project ('Sustainable Urban Water Management Improves Tomorrow's City's Health'), supported by the European Commission under the 6th Framework Programme. In the SWITCH project context, Alberto Galvis started his doctoral research. This investigation was completed with the support of the project 'Strategies for recovery and integrated water resources management in the Cauca and Dagua basins, in the Cauca Valley', implemented by the Universidad del Valle and supported by the Sistema Nacional de Regalias and the Government of Valle del Cauca, Colombia.

Acknowledgements of financial support

This research was carried out with support of the SWITCH project ('Sustainable Urban Water Management Improves Tomorrow's City's Health'). SWITCH was supported by the European Commission under the 6th Framework Programme, and contributing to the thematic priority area of 'Global Change and Ecosystems' [1.1.6.3] Contract N° 018530-2. In its final stage the research received support from the project 'Strategies for recovery and integrated water resources management in the Cauca and Dagua basins, in the Cauca Valley', implemented by Universidad del Valle and funded by 'Sistema Nacional de Regalias' and Government of Valle del Cauca, Colombia.